Electrical Wiring Residential Lab Manual

18th Edition

Walter C. Bartlett, Ph.D.

Australia • Brazil • Japan • Korea • Mexico • Singapore • Spain • United Kingdom • United States

Electrical Wiring Residential Lab Manual, 18E
Walter C. Bartlett, Ph.D.

Vice President, GM Skills & Product Planning:
Dawn Gerrain

Product Team Manager: James DeVoe

Senior Director, Development: Marah
Bellegarde

Senior Product Development Manager:
Larry Main

Senior Product Manager: John Fisher

Product Assistant: Andrew Ouimet

Vice President, Marketing Services:
Jennifer Baker

Marketing Manager: Linda Kuper

Senior Production Director: Wendy A. Troeger

Production Manager: Mark Bernard

Senior Content Project Manager:
Kara A. DiCaterino

Senior Art Director: David Arsenault

Cover Images: © iStockphoto/piovesempre

© iStockphoto/tarczas

For product information and technology assistance, contact us at
Cengage Learning Customer & Sales Support, 1-800-354-9706

For permission to use material from this text or product,
submit all requests online at **www.cengage.com/permissions**.
Further permissions questions can be e-mailed to
permissionrequest@cengage.com

ISBN-13: 978-1-285-17112-8

ISBN-10: 1-285-17112-8

Cengage Learning
200 First Stamford Place, 4th Floor
Stamford, CT 06902
USA

Cengage Learning is a leading provider of customized learning solutions with office locations around the globe, including Singapore, the United Kingdom, Australia, Mexico, Brazil, and Japan. Locate your local office at:
www.cengage.com/global

Cengage Learning products are represented in Canada by Nelson Education, Ltd.

To learn more about Cengage Learning, visit **www.cengage.com**

Purchase any of our products at your local college store or at our preferred online store **www.cengagebrain.com**

Notice to the Reader
Publisher does not warrant or guarantee any of the products described herein or perform any independent analysis in connection with any of the product information contained herein. Publisher does not assume, and expressly disclaims, any obligation to obtain and include information other than that provided to it by the manufacturer. The reader is expressly warned to consider and adopt all safety precautions that might be indicated by the activities described herein and to avoid all potential hazards. By following the instructions contained herein, the reader willingly assumes all risks in connection with such instructions. The publisher makes no representations or warranties of any kind, including but not limited to, the warranties of fitness for particular purpose or merchantability, nor are any such representations implied with respect to the material set forth herein, and the publisher takes no responsibility with respect to such material. The publisher shall not be liable for any special, consequential, or exemplary damages resulting, in whole or part, from the readers' use of, or reliance upon, this material.

Printed in the United States of America
1 2 3 4 5 6 7 18 17 16 15 14

Contents

Preface

The primary audience for this laboratory manual is post-secondary electrical wiring students preparing to work in the field of residential electricity. However, due to the student-friendly design of the lab exercises, as well as lab sequence flexibility, this laboratory manual could be utilized in high school house wiring programs and in training divisions of companies requiring personnel with residential electricity skills. It was written to accompany *Electrical Wiring—Residential* by Ray Mullin and Phil Simmons, but it can be used in any residential wiring lab program. The main objective of this laboratory manual is to take a novice from simplistic wiring circuits to progressively harder and more complicated wiring circuitry. Furthermore, the labs are designed to provide students not only with practical hands-on wiring experience but also with an opportunity to read, study, and become familiar with the *National Electrical Code®* (*NEC®*)* codes and regulations that pertain to the particular lab assignments.

When I started teaching a residential wiring class as an instructor in the North Carolina Community College System, it soon became clear to me that there were no residential wiring laboratory manuals available. I have taught numerous electronic and electrical courses during my past twenty-seven years of teaching, and I have found that a laboratory manual is invaluable in any lab-based course. The residential wiring course is not unlike any other lab-based course in that the students would benefit greatly from having a laboratory manual that they could follow, take home, and study. Therefore, I decided I would produce the needed laboratory manual that could be used to provide entry-level electricians with the fundamental wiring skills and hands-on experience, using the electrical circuit theories associated with residential wiring. This manual would also be diverse enough to be used in a variety of electrical training programs and environments, for example, high school vocational courses, community college curricula, and/or continuing education courses, and industrial training sessions.

Lab Structure

This laboratory manual utilizes an incremental approach to teaching residential wiring circuits, by starting with a basic lighting circuit controlled at one location and advancing to more complex residential wiring circuits. Each lab consists of two main sections: a hands-on wiring exercise and *NEC* drill problems. In each lab the student draws the floor plan of the wiring booth, using the appropriate electrical symbols, and then draws the actual wiring connections needed to wire the assigned wiring circuit. Once the student has finished drawing the actual wiring connections, the instructor verifies the drawing's validity and initials the lab assignment, so the student may begin wiring the electrical circuit. The instructor verifies that the circuit has been wired and operates properly, and again initials the student's lab sheet. The student secures the assigned wiring booth and then completes the *NEC* Drill Work section of the lab assignment.

Notable Features

Consistency: The layout of labs is consistent throughout the lab manual so that students become familiar with the structure of the lab assignments. This helps focus students' time and effort on the assigned wiring circuitry rather than on trying to understand a different set of directions and format for each lab task.

Floor Plan Drawing: Each lab experience requires students to draw the electrical layout of their respective booths in accordance with the lab assignment. Drawing the electrical layout allows students the opportunity to become familiar with basic electrical symbols.

---

*National Electrical Code® and *NEC®* are registered trademarks of the National Fire Protection Association, Inc., Quincy, MA *02269.*

Actual Wiring Diagram: Students must draw their plan to make the electrical connections for the assigned lab, and once the instructor has verified the drawing's validity, students construct the wiring circuit utilizing their electrical drawing as a guide. This serves as a simulation of tasks students will be required to perform in their future workplaces.

NEC Drill Work: Each lab is concluded with a series of questions that require the students to spend time researching the answers using the *NEC* so that they become more and more familiar with the *NEC*. This is a significant piece in preparing for the workplace, which requires a thorough familiarity with the *NEC*.

Instructor Verifications: At strategic points during the lab experience, the instructor is to verify the students' work and provide them with either positive feedback or take the opportunity to perform one-on-one instruction. This allows the instructor to either offer immediate positive feedback or lend guidance to the students on a one-on-one basis. This method provides students with ongoing assessment and immediate reinforcement rather than delayed feedback that allows repetition of error and its consequent reinforcement.

Lab Sequence: The sequence of the wiring assignments is only a suggestion; therefore, the labs can readily be assigned out of sequence so that the instructor can fit the labs to the lesson being taught at the time. This adds significantly to the manual's flexibility for use in a variety of electrical training programs.

Self-Assessment of Lab Activity: Each lab assignment has students reflect on what they have just learned. This time of self-assessment benefits both the students and the instructor. The students benefit because as they spend time reflecting on what they have just learned, they can formulate questions that they still need to ask the instructor about the lab assignment. In turn, the instructor can use this information to help determine what material was well taught and what material needs to be reviewed.

Acknowledgments

Cengage Learning and the author gratefully acknowledge the time and suggestions put forth by the review panel. Our special thanks to:

DeWain Belote
Hillsborough County Community College
Tampa, FL

Keith Bunting
Randolph Community College
Asheboro, NC

Keith Elliott
Rockingham Community College
Westworth, NC

Daniel Lewis
James Rumsey Technical Institute
Martinsburg, WV

Debra Matthews
Mississippi Gulf Coast Community College
Gautier, MS

Gary Reiman
Dunwoody Technical Institute
Minneapolis, MN

Bill Whitcomb
Independent Electrical Contractors
Denver, CO

LAB EXERCISE #1

Electrical Safety

Name:	Date:	Grade:

Student Learning Outcomes

After completing this lab, students should be able to:

- Identify the three major categories of electrical safety.
- Identify a minimum of seven safety rules relating to personnel safety.
- Identify a minimum of four safety rules relating to equipment safety.
- Identify a minimum of five safety rules relating to work area safety.

Equipment and Supplies

Three sheets of transparencies per group

Two different-colored transparency markers per group

Overhead projector

Procedures

1. The class is to be divided into three groups, and each group must elect a note taker or recorder, a reporter, and a facilitator for large-group discussion.

2. Each group uses one transparency to record the names of the members of the group, which are used during the group's presentation.

3. Each group is assigned one of the following topics to brainstorm and presents its findings to the rest of the class.
   A. Personnel safety rules
   B. Equipment safety rules
   C. Work area safety rules

4. During the brainstorming session, the members of each group identify as many safety rules as they can that should be adhered to for the safety category the group was assigned. These rules should be written on a transparency sheet that will be used during the group's presentation.

5. After an allotted amount of time determined by the instructor, the reporter of each group presents the group's findings to the rest of the class. Space has been provided to document the names of the members and the safety rules presented by each group.

6. After the group reporter has presented the group's findings, the group facilitator asks the rest of the class whether there are any other rules that they feel should be included, and the facilitator adds these suggestions to the group's findings.

Personnel Safety Rules

Group Members: _____   _____

                       _____   _____

                       _____   _____

                       _____   _____

Identify and record any safety rules that pertain to the safekeeping of personnel working in the electrical field.

Rule #1 _____

Rule #2 _____

Rule #3 _____

Rule #4 _____

Rule #5 _____

Rule #6 _____

Rule #7 _____

Rule #8 _____

Rule #9 _____

Rule #10 _____

Rule #11 _____

Equipment Safety Rules

Group Members: _____   _____

                       _____   _____

                       _____   _____

                       _____   _____

Identify and record any safety rules that help ensure that the equipment an electrician is working with (or on) is not damaged during general operation and/or maintenance procedures.

Rule #1 _____

Rule #2 _____

Rule #3 _____

Rule #4 _____

Rule #5 _____

Rule #6 _____

Rule #7 _____

Rule #8 _____

Rule #9 _____

Rule #10 _____

Rule #11 _____

Work Area Safety Rules

Group Members: _____   _____

                  _____   _____

                  _____   _____

                  _____   _____

Identify and record any safety rules required in maintaining a safe work area.

Rule #1 _____

Rule #2 _____

Rule #3 _____

Rule #4 _____

Rule #5 _____

Rule #6 _____

Rule #7 _____

Rule #8 _____

Rule #9 _____

Rule #10 _____

Rule #11 _____

7. Electricians must be aware of their environment and pay special attention to the first-aid and emergency items located in their work area. Find and document the location of the items listed here.

   A. Fire extinguisher: _____

   B. First-aid kit: _____

   C. Master power switch: _____

   D. Fire exits: _____

   E. Fire alarm switches: _____

8. In the space provided here, draw the general floor plan of the lab facility, and identify the location of the first-aid and emergency items listed in the previous step.

9. What did you find was the most challenging part of this lab assignment?

   _____

   _____

   _____

10. What was the most interesting or important thing that you learned during this lab assignment?

   _____

   _____

   _____

National Electrical Code® Drill Work

Select the correct answer for the following questions, and indicate the article or section in the *NEC*® in which the answer can be found.

1. How often is the *NEC* revised and republished?
   A. Every 3 months
   B. Every 2 years
   C. Every 3 years
   D. Every 5 years

2. How are changes in the 2011 edition of the *NEC* (compared with those in the 2008 edition) identified?
   A. By an asterisk
   B. By a bullet   *NEC*: _____
   C. With a vertical line in the margin
   D. By highlighting with gray shading

3. True or false? The purpose of the *NEC* is to accommodate future expansion of electrical use.
   A. True
   B. False   *NEC*: _____

4. The *NEC* does not cover any of the following, *except*:
   A. Installations in ships
   B. Installations of optical fiber raceways   *NEC*: _____
   C. Installations underground in mines
   D. Installation of railways for generation of power used exclusively for operation of rolling stock

5. Which article of the *NEC* deals specifically with lighting systems operating at 30 volts or less?
   A. *220*
   B. *411*
   C. *517*
   D. *553*

6. By which of the following organizations is *NEC* sponsored?
   A. National Fire Protection Association
   B. Society of Manufacturing Engineers
   C. IEEE
   D. National Science Foundation

Pre-Course Knowledge Assessment Lab

Name:	Date:	Grade:

Student Learning Outcomes

After completing this lab, students should be able to:

- State which components and hardware of residential electricity they currently understand.

- Recognize which components and hardware of residential electricity they need to master before the conclusion of this course.

- State which residential electricity tools and equipment they currently understand.

- Recognize which residential electricity tools and equipment they need to master prior to the conclusion of this course.

Equipment and Supplies

The instructor secures 25–30 commonly used residential wiring parts and components and 10–15 commonly used hand tools and power tools that a practicing electrician needs to know how and when to use.

Procedures

1. The instructor has laid out 25–30 commonly used residential parts and components, which you will try to identify.

2. Identify the name of each electrical component, and write a brief description of what it is and how the component is used in residential wiring.

Part #1: _____ : _____

_____

Part #2: _____ : _____

_____

Part #3: _____ : _____

_____

Part #4: _____ : _____

_____

Part #5: _____ : _____

_____

Part #6: _____ : _____

_____

Part #7: _____ : _____

_____

Part #8: _____ : _____

_____

Part #9: _____ : _____

_____

Part #10: _____ : _____

_____

Part #11: _____ : _____

_____

Part #12: _____ : _____

_____

Part #13: _____ : _____

_____

Part #14: _____ : _____

_____

Part #15: _____ : _____

_____

Part #16: _____ : _____

_____

Part #17: _____ : _____

_____

Part #18: _____ : _____

_____

Part #19: _____ : _____

_____

Part #20: _____ : _____

_____

Part #21: _____ : _____

_____

Part #22: _____ : _____

_____

Part #23: _____ : _____

_____

Part #24: _____ : _____

_____

Part #25: _____ : _____

_____

Part #26: _____ : _____

_____

Part #27: _____ : _____

_____

Part #28: _____ : _____

_____

Part #29: _____ : _____

_____

Part #30: _____ : _____

_____

3. The instructor has laid out 10–15 commonly used residential wiring hand tools and power tools, which you will try to identify.

4. Identify the name of each tool, and write a brief description of how the tool is used in residential wiring.

Tool #1: _____ : _____
_____

Tool #2: _____ : _____
_____

Tool #3: _____ : _____
_____

Tool #4: _____ : _____
_____

Tool #5: _____ : _____
_____

Tool #6: _____ : _____
_____

Tool #7: _____ : _____
_____

Tool #8: _____ : _____
_____

Tool #9: _____ : _____
_____

Tool #10: _____ : _____
_____

Tool #11: _____ : _____
_____

Tool #12: _____ : _____
_____

Tool #13: _____ : _____
_____

Tool #14: _____ : _____

_____

Tool #15: _____ : _____

_____

5. How will this lab assignment affect the way you proceed in this class?

_____

_____

6. What was the most interesting or important thing that you learned during this lab assignment?

_____

_____

National Electrical Code® Drill Work

Select the correct answer for each of the following questions, and indicate the article or section in the *NEC*® in which the answer can be found.

1. What is the maximum distance from the outside edge of a bathroom basin that a wall receptacle outlet can be located within dwelling units?
   A.   3 ft (914 mm)
   B.   4 ft (1.2 m)   *NEC*: _____
   C.   5 ft (1.5 m)
   D.   6 ft (1.8 m)

2. When referring to circuit breakers, what is the qualifying term indicating that there is a purposely introduced delay in the tripping action of the circuit breaker, which delay decreases as the magnitude of the current increases?
   A.   Instantaneous trip
   B.   Direct time   *NEC*: _____
   C.   Inverse time
   D.   Rapid response time

3. Which of the following articles deals specifically with the installation of fixed electric space-heating equipment?
   A.   *312*
   B.   *210*
   C.   *410*
   D.   *424*

4. True or false? It is highly recommended, but not required, that all branch circuits that supply 125-volt, single-phase, 15- and 20-ampere outlets installed in dwelling unit bedrooms be protected by an arc-fault circuit interrupter.
   A.   True
   B.   False   *NEC*: _____

5. According to the *NEC*, the metric designators for the trade sizes 1 and 4 are ___ and ___, respectively.

   A. 12 and 27

   B. 21 and 53   *NEC*: _____

   C. 27 and 103

   D. 78 and 103

6. So that required examinations, adjustments, servicing, and/or maintenance can be accomplished safely, the minimum width of a working space allowed for equipment operating at 600 volts nominal or less to ground must be equal to the width of the equipment or ___, whichever is greater.

   A. 2.5 ft (762 mm)

   B. 3 ft (914 mm)   *NEC*: _____

   C. 4 ft (1.2 m)

   D. 6 ft (1.8 m)

LAB EXERCISE #3

Conductor Stripping and Connection Techniques

Name:	Date:	Grade:

Student Learning Outcomes

After completing this lab, students should be able to:

- Strip electrical conductors and cables, using various wire-stripping techniques and tools.
- Secure multiple conductors together using the twist-on wire connection techniques.
- Splice various size conductors together using solderless terminals.
- Use twist-on grounding wire connectors.
- Fasten conductors together using crimp sleeve connectors.

Equipment and Supplies

10 AWG insulated solid conductor

10 AWG insulated stranded conductor

12 AWG insulated solid conductor

12 AWG insulated stranded conductor

14 AWG insulated solid conductor

16 AWG insulated solid conductor

14/2 NMB cable

Various sizes twist-on wire connectors

Solderless terminal connectors for 10, 12, and 14 AWG conductors

Grounding connection terminals: twist-on grounding connectors and crimp sleeve connectors

Procedures

Conductor Stripping Techniques

1. Obtain the following items from the instructor:
   - A 1 ft (305 mm) length of 10 AWG solid and stranded conductors
   - A 1 ft (305 mm) length of 12 AWG solid and stranded conductors
   - A 2 ft (610 mm) section of 14/2 NMB cable

2. Using an electrician's knife, remove ¾ in. (19.05 mm) of insulation from one end of each of the 10 AWG and 12 AWG solid and stranded conductors.

13

3. Using an electrician's knife, remove 1 ft (305 mm) of the NMB cable's outer sheathing so that 1 ft (305 mm) of the two insulated conductors and the bare ground conductor are exposed.

4. Using an electrician's knife, remove ¾ in. (19.05 mm) of the insulation from the ungrounded and grounded conductors.

5. Using long-nose pliers, bend the ends of the four insulated conductors and the three conductors of the NMB cable into loops that will surround a potential terminal screw. The conductor loops should be bent to form between 240° and 270° of a complete circle. Once these tasks have been completed, have the instructor inspect the conductor loops and initial here when these tasks have been mastered.

Instructor's initials: _____

6. Using lineman's pliers, remove the stripped portions of the four conductors and the NMB cable by cutting off 1 in. (25.4 mm) of the conductors.

7. Using basic wire strippers/cutters, remove ¾ in. (19.05 mm) of insulation from one end of the 10 AWG and 12 AWG solid and stranded conductors.

8. Using basic wire strippers/cutters, remove ¾ in. (19.05 mm) of the insulation from the ungrounded and grounded conductors of the 14/2 NMB cable.

9. Using the appropriate hole of the basic wire strippers/cutters, bend the ends of the four insulated conductors and the three conductors of the NMB cable into loops that will surround a potential terminal screw. The conductor loops should be bent to form between 240° and 270° of a complete circle. Once these tasks have been completed, have the instructor inspect the conductor loops and initial here when these tasks have been mastered.

Instructor's initials: _____

10. Using lineman's pliers, remove the stripped portions of the four conductors and the NMB cable by cutting off 1 in. (25.4 mm) of the conductors.

11. Using automatic wire strippers, remove ¾ in. (19.05 mm) of insulation from one end of each of the 10 AWG and 12 AWG solid and stranded conductors.

12. Using automatic wire strippers, remove ¾ in. (19.05 mm) of the insulation from the ungrounded and grounded conductors of the 14/2 NMB cable. Once these tasks have been completed, have the instructor inspect the conductors and initial here when these tasks have been mastered.

Instructor's initials: _____

Twist-On Wire Connection Techniques

13. Use the proper twist-on wire connectors to secure the following conductor connections. Once you have completed all four conductor connection combinations, have the instructor inspect the connections and initial here after each connection that was accomplished correctly.

Conductor Connections	Twist-On Wire Connector Color	Instructor's Initials
Two 16 AWG	_____	_____
Four 12 AWG	_____	_____
One 14 AWG and one 16 AWG	_____	_____
Two 10 AWG and one 12 AWG	_____	_____

Solderless Terminal Connection Techniques

14. Use the proper solderless terminals to make three connections on 14 AWG conductors. Once these connections have been completed, have the instructor inspect the splices and initial here when these tasks have been mastered.

Instructor's initials: _____

15. Use the proper solderless terminals to make three connections on 12 AWG conductors. Once these connections have been completed, have the instructor inspect the splices and initial here when these tasks have been mastered.

Instructor's initials: _____

16. Use the proper solderless terminals to make three connections on 10 AWG conductors. Once these connections have been completed, have the instructor inspect the splices and initial here when these tasks have been mastered.

Instructor's initials: _____

Grounding Connection Techniques

17. Use the proper twist-on grounding connector to secure the grounding conductors of two 14/2 NMB cables to the grounding screw of a metal device box. Once this task has been completed, have the instructor inspect the connection and initial here when this task has been mastered.

Instructor's initials: _____

18. Properly strip and twist three 14 AWG conductors together, and use a crimp sleeve connector to secure the conductors together, with one conductor long enough to fasten to the grounding screw of a metal device box. Once this task has been completed, have the instructor inspect the connection and initial here when this task has been mastered.

Instructor's initials: _____

19. Properly strip and twist three 12 AWG conductors together, and use a crimp sleeve connector to secure the conductors together with one conductor long enough to fasten to the grounding screw of a metal device box. Once this task has been completed, have the instructor inspect the connection and initial here when this task has been mastered.

Instructor's initials: _____

20. What did you find was the most challenging part of this lab assignment?

_____

_____

_____

21. What was the most interesting or important thing that you learned during this lab assignment?

_____

_____

_____

National Electrical Code® Drill Work

Select the correct answer for the following questions and indicate the article or section in the *NEC*® in which the answer can be found.

1. How is the deletion of complete paragraphs identified in the 2014 *NEC*?
   A. By an asterisk located in the margin
   B. By a bullet placed between the paragraphs that remain   NEC: _____
   C. By a vertical line in the left margin
   D. By a vertical line in the right margin

2. The *NEC* is divided into the following sequence:
   A. Articles, chapters, sections, subsections, and exceptions
   B. Sections, chapters, articles, subsections, and exceptions
   C. Chapters, articles, sections, subsections, and exceptions
   D. None of the above

3. True or false? The NFPA has no power to police or enforce compliance with the contents of the *NEC*.
   A. True
   B. False   NEC: _____

4. Which article of the *NEC* deals with the installation and construction specifications of nonmetallic sheathed cable?
   A. *340*
   B. *384*
   C. *530*
   D. *334*

5. True or false? It is required that all branch circuits that supply 125-volt, single-phase, 15- and 20-ampere outlets installed in dwelling unit hallways shall be protected by a listed arc-fault circuit interrupter.
   A. True
   B. False   NEC: _____

6. The *NEC* covers the following subjects, *except*:
   A. Installation of equipment that connects to the supply of electricity
   B. Installation of fiber-optic cable   NEC: _____
   C. Installation of electric conductors within private buildings
   D. Installations of communications equipment under the exclusive control of communications utilities located outdoors

Single-Pole Switched Light Circuit with Feed at the Switch

Name:	Date:	Grade:

Student Learning Outcomes

After completing this lab, students should be able to:

- Perform the calculations involved in selecting the correctly sized boxes for the assigned task.

- Properly mount the boxes needed for a lighting circuit that is controlled at one location, with the 120-volt source entering at the switch.

- Properly install the wiring and connect the devices required for a lighting circuit that is controlled at one location, with the 120-volt source entering at the switch.

Equipment and Supplies

Single-pole switch, switch box

120-volt luminaire, lighting outlet

14/2 NMB cable

Cable staples

Procedures

> **CAUTION**
> Because portions of this lab involve working with 120 vac, your complete attention is required, and your every action must be well thought out and safety conscious.

> **POWER TOOL CAUTION**
> During the course of this lab assignment, it may be necessary to use various power tools and stepladders. It is imperative that a thorough understanding of the operating procedures and safety concerns associated with these power tools and stepladders be achieved before using them.

1. In the space provided on the next page, sketch the basic floor plan of your wiring booth, and draw the wiring diagram needed to wire a lighting circuit that is controlled at one location, with the 120-volt source entering at the switch. (Refer to Appendix I for an example floor plan.) The circuit should be laid out in the following manner:

A. The switch should be located at the front of the left-hand panel and 4 ft (1.2 m) from the floor to the bottom of the box.

B. The luminaire should be located at the front of the right-hand panel and 6 ft (1.8 m) from the floor to the center of the box.

C. The wiring should pass overhead.

2. Calculate the minimum cubic-inch volume required for each of the boxes installed (using 14/2 NMB cable), and record your results here.

Switch box   =   _____

Luminaire   =   _____

3. Indicate in the spaces here the type and the minimum capacity (in.^{3}) of the boxes used in this lab (refer to *Table 314.16(A)* of the *NEC®*) (1 in.^{3} = 16.4 cm^3).

		Type of Box	**Minimum Capacity (in.^{3})**
Switch box	=	_____	_____
Luminaire	=	_____	_____

4. What is the minimum length of conductor required from where the conductor emerges from the cable or raceway, or enters the outlet, to the end of the conductor? Where in the *NEC* is this stated?

_____

5. What is the maximum distance that is allowed between the box and the first cable strap or staple? Where in the *NEC* is this stated?

_____

6. What is the maximum distance that is allowed between cable straps or staples? Where in the *NEC* is this stated?

_____

7. What is the minimum radius allowed along the inner edge of a bent cable? Where in the *NEC* is this stated?

_____

8. In the space provided here, draw the exact wiring that is involved in a lighting circuit that is controlled at one location, with the 120-volt source entering at the switch. (Refer to Appendix II for an example wiring diagram.) Once this task has been completed, have the instructor verify that your diagram is drawn correctly and initial below.

CAUTION

It is important that the instructor check your wiring diagram before you begin to complete the electrical connections.

Instructor's initials: _____

9. Once the instructor has verified the validity of your wiring diagram, wire your assigned booth in accordance with the wiring diagram in step 8 (account for ½ in. [12.7 mm] thick gypsum board).

CAUTION

Do not work on circuits with power applied. Notify the instructor before energizing the circuit, and de-energize the power whenever possible.

---

10. Once you have completed the installation, have the instructor inspect and grade your wiring booth, using the following criteria:

Electrical integrity	_____	(Splices, terminations, and grounding were correctly done.)
NEC compliance	_____	(Local codes should be included.)
Operational	_____	(The circuit operates as defined.)
Appearance	_____	(Work shows a neat and professional manner.)
Safety factor	_____	(Student used proper safety practices and operated tools and equipment safely.)

CAUTION

Be sure to de-energize the electrical circuit before disassembling the wiring booth.

---

11. What did you find was the most challenging part of this lab assignment?

_____

_____

_____

12. What was the most interesting or important thing that you learned during this lab assignment?

_____

_____

_____

National Electrical Code® Drill Work

Use *Article 100* of the *NEC* to determine which term is being defined.

1. Connected to ground or to a conductive body that extends the ground connection

_____

2. A point on the wiring system at which current is taken to supply utilization equipment

_____

3. Constructed so that moisture will not enter the enclosure under specified test conditions

_____

4. The circuit conductors between the final overcurrent device protecting the circuit and the outlet(s)

_____

5. A unit of an electrical system, other than a conductors, that carries or controls electric energy as its principal function

_____

Use *Article 100* of the *NEC* to define the following terms.

6. Overcurrent:

_____

_____

7. Ampacity:

_____

_____

8. Service point:

_____

_____

9. Bonding conductor or jumper:

_____

_____

10. Feeder:

_____

_____

Single-Pole Switched Light Circuit with Feed at the Light

Name:	Date:	Grade:

Student Learning Outcomes

After completing this lab, students should be able to:

- Perform the calculations involved in selecting the correctly sized boxes for the assigned task.

- Properly mount the boxes needed for a lighting circuit that is controlled at one location, with the 120-volt source entering at the lighting outlet.

- Properly install the wiring and connect the devices required for a lighting circuit that is controlled at one location, with the 120-volt source entering at the lighting outlet.

Equipment and Supplies

Single-pole switch, switch box

120-volt luminaire, lighting outlet

14/2 NMB cable

Cable staples

Procedures

> **CAUTION**
> Because portions of this lab involve working with 120 vac, your complete attention is required, and your every action must be well thought out and safety conscious.

> **POWER TOOL CAUTION**
> During the course of this lab assignment, it may be necessary to use various power tools and stepladders. It is imperative that a thorough understanding of the operating procedures and safety concerns associated with these power tools and stepladders be achieved before using these tools and ladders.

1. In the space provided on the next page, sketch the basic floor plan of your wiring booth, and draw the wiring diagram that is needed to wire a lighting circuit that is controlled at one location, with the 120-volt source entering at the lighting outlet. (Refer to Appendix I for an example floor plan.) The circuit should be laid out in the following manner:

A. The switch should be located at the front of the left-hand panel and 4 ft (1.2 m) from the floor to the bottom of the box.

B. The luminaire should be located at the front of the right-hand panel and 6 ft (1.8 m) from the floor to the center of the box.

C. The wiring should pass overhead.

2. Calculate the minimum cubic-inch volume required for each of the boxes installed, and record your results here.

Switch box   =   _____

Luminaire   =   _____

3. Indicate in the spaces here the type and the minimum capacity (in.³) of the boxes used in this lab (refer to *Table 314.16(A)* of the *NEC®*) (1 in.³ = 16.4 cm³).

		Type of Box	**Minimum Capacity (in.³)**
Switch box	=	_____	_____
Luminaire	=	_____	_____

4. What is the maximum distance that is allowed between cable straps or staples? Where in the *NEC* is this stated?

_____

5. What is the minimum length of conductor required from where the conductor emerges from the cable or raceway, or enters the outlet, to the end of the conductor? Where in the *NEC* is this stated?

_____

6. What is the minimum radius allowed along the inner edge of a bent cable? Where in the *NEC* is this stated?

_____

7. What is the maximum distance that is allowed between the box and the first cable strap or staple? Where in the *NEC* is this stated?

   _____

8. In the space provided here, draw the exact wiring that is involved in a lighting circuit that is controlled at one location, with the 120-volt source entering at the lighting outlet. (Refer to Appendix II for an example wiring diagram.) Once this task has been completed, have the instructor verify that your diagram is drawn correctly and initial below.

CAUTION

It is important that the instructor check your wiring diagram before you begin to complete the electrical connections.

Instructor's initials: _____

9. Once the instructor has verified the validity of your wiring diagram, wire your assigned booth in accordance with the wiring diagram in step 8 (account for ½ in. [12.7 mm] thick gypsum board).

CAUTION

Do not work on circuits with power applied. Notify the instructor before energizing the circuit, and de-energize the power whenever possible.

---

10. Once you have completed the installation, have the instructor inspect and grade your wiring booth, using the following criteria:

   Electrical integrity   _____   (Splices, terminations, and grounding were correctly done.)

   NEC compliance   _____   (Local codes should be included.)

   Operational   _____   (The circuit operates as defined.)

   Appearance   _____   (Work shows a neat and professional manner.)

   Safety factor   _____   (Student used proper safety practices and operated tools and equipment safely.)

CAUTION

Be sure to de-energize the electrical circuit before disassembling the wiring booth.

---

11. What did you find was the most challenging part of this lab assignment?

_____

_____

_____

12. What was the most interesting or important thing that you learned during this lab assignment?

_____

_____

_____

National Electrical Code® Drill Work

Use *Article 110* of the *NEC* to answer the following questions and indicate the exact section number of your answer.

1. How is electrical equipment to be mechanically installed?

_____

_____

2. How are conductor sizes expressed?

_____

_____

3. What is stated about indoor electrical installations that are over 600 volts and are accessible to unqualified persons?

_____

_____

4. What is stated about electrical equipment that produces arcs and sparks?

_____

_____

5. What is stated about unused openings in boxes and raceways?

_____

_____

6. What does the *Code* state concerning the manufacturer's markings of electrical equipment?

_____

_____

Dimmer Switched Light Circuit and Continuously Energized Receptacle

Name:	Date:	Grade:

Student Learning Outcomes

After completing this lab, students should be able to:

- Perform the calculations involved in selecting the correctly sized boxes for the assigned task.

- Properly mount the boxes needed for a lighting circuit that is controlled at one location, with the 120-volt source entering at a dimmer switch and the duplex receptacle being continuously energized.

- Properly install the wiring and connect the devices required for a lighting circuit that is controlled at one location, with the 120-volt source entering at a dimmer switch and the duplex receptacle being continuously energized.

Equipment and Supplies

Dimmer switch, switch box

120-volt luminaire, lighting outlet

Duplex receptacle, receptacle outlet

14/2 NMB cable

14/3 NMB cable

Cable staples

Procedures

CAUTION
Because portions of this lab involve working with 120 vac, your complete attention is required, and your every action must be well thought out and safety conscious.

POWER TOOL CAUTION
During the course of this lab assignment, it may be necessary to use various power tools and stepladders. It is imperative that a thorough understanding of the operating procedures and safety concerns associated with these power tools and stepladders be achieved before using them.

1. In the space provided here, sketch the basic floor plan of your wiring booth, and draw the wiring diagram that is needed to wire a lighting circuit that is controlled at one location, with the 120-volt source entering at a dimmer switch and the receptacle being continuously energized. (Refer to Appendix I for an example floor plan.) The circuit should be laid out in the following manner:

A. Dimmer switch should be located at the front of the right-hand panel and 4 ft (1.2 m) from the floor to the bottom of the box.

B. Luminaire should be located overhead and in the center of the booth.

C. The duplex receptacle should be located at the front of the left-hand panel and 1 ft (305 mm) from the floor to the bottom of the outlet.

2. Calculate the minimum cubic-inch volume required for each of the boxes installed (using the appropriate 14 AWG NMB cable), and record your results here.

Switch box   =   _____

Lighting outlet   =   _____

Receptacle outlet   =   _____

3. Indicate in the spaces here the type and the minimum capacity (in.³) of the boxes used in this lab (refer to *Table 314.16(A)* of the *NEC®*) (1 in.³ = 16.4 cm³).

		Type of Box	**Minimum Capacity (in.³)**
Switch box	=	_____	_____
Lighting outlet	=	_____	_____
Receptacle outlet	=	_____	_____

4. What are the acceptable colors of a 12 AWG grounded conductor? Where in the *NEC* is this stated?

_____

_____

5. What is the color of the grounded conductor that is being used to wire your electrical booth?

_____

_____

6. In the space provided here, draw the exact wiring that is involved in a lighting circuit that is controlled at one location, with the 120-volt source entering at a dimmer switch and the receptacle being continuously energized. (Refer to Appendix II for an example wiring diagram.) Once this task has been completed, have the instructor verify that your diagram is drawn correctly and initial below.

> **CAUTION**
>
> It is important that the instructor check your wiring diagram before you begin to complete the electrical connections.

Instructor's initials: _____

7. Once the instructor has verified the validity of your wiring diagram, wire your assigned booth in accordance with the wiring diagram in step 6 (account for ½ in. [12.7 mm] thick gypsum board).

> **CAUTION**
>
> Do not work on circuits with power applied. Notify the instructor before energizing the circuit, and de-energize the power whenever possible.

8. Once you have completed the installation, have the instructor inspect and grade your wiring booth, using the following criteria:

Electrical integrity   _____   (Splices, terminations, and grounding were correctly done.)

NEC compliance   _____   (Local codes should be included.)

Operational   _____   (The circuit operates as defined.)

Appearance   _____   (Work shows a neat and professional manner.)

Safety factor   _____   (Student used proper safety practices and operated tools and equipment safely.)

> **CAUTION**
>
> Be sure to de-energize the electrical circuit before disassembling the wiring booth.

9. What did you find was the most challenging part of this lab assignment?

_____

_____

_____

10. What was the most interesting or important thing that you learned during this lab assignment?

_____

_____

_____

National Electrical Code® Drill Work

Use the *NEC* to determine which answer is correct, and indicate the article or section in the *NEC* in which the answer can be found.

1. Which statement(s) about the inspection of electrical work is (are) correct?

   I. Inspectors shall make as many inspections as necessary to ensure compliance with applicable laws.
   II. An inspector shall issue a certificate of compliance when the completed electrical installation complies with all applicable laws and with the terms of the permit.
   A.   I only
   B.   II only
   C.   Both I and II
   D.   Neither I nor II

2. In dwelling units, ground-fault circuit interrupter protection is required in all of the following locations, *except*:
   A.   Garages
   B.   Outdoors   *NEC*: _____
   C.   Bedrooms
   D.   Bathrooms

3. What is the minimum number of 120-volt, 15-ampere, 2-wire lighting branch circuits required for a residence that is 65 ft × 35 ft (19.8 m × 10.67 m)?
   A.   2
   B.   3
   C.   4
   D.   5

4. An individual 20-ampere, rated branch circuit serves a single receptacle. The rating of the receptacle must not be less than:
   A.   12 amperes
   B.   15 amperes   *NEC*: _____
   C.   16 amperes
   D.   20 amperes

5. The grounded conductor of a branch circuit shall be identified by an outer covering color of any of the following, *except*:

   A. Gray

   B. White

   C. Yellow with a green stripe

   D. Blue with three white stripes

   NEC: _____

6. The rating of the branch-circuit overcurrent device supplying a low-voltage wiring system is 15 amperes. What is the minimum allowable size equipment grounding conductor required for this dedicated circuit?

   A. 14 AWG copper

   B. 14 AWG aluminum

   C. 12 AWG copper

   D. 10 AWG aluminum

   NEC: _____

Single-Pole Switched Light Circuit and Two Split Duplex Receptacles

Name:	Date:	Grade:

Student Learning Outcomes

After completing this lab, students should be able to:

- Perform the calculations involved in selecting the correctly sized boxes for the assigned task.

- Properly mount the boxes needed for a lighting circuit that is controlled at one location, with the 120-volt source entering at the switch and two duplex receptacles that are split, with half of each receptacle controlled by the switch.

- Properly install the wiring and connect the devices required for a lighting circuit that is controlled at one location, with the 120-volt source entering at the switch and two duplex receptacles that are split, with half of each receptacle controlled by the switch.

Equipment and Supplies

Single-pole switch, switch box

120-volt luminaire, lighting outlet

Two duplex receptacles, receptacle outlets

14/2 NMB cable

14/3 NMB cable

Cable staples

Procedures

CAUTION
Because portions of this lab involve working with 120 vac, your complete attention is required, and your every action must be well thought out and safety conscious.

POWER TOOL CAUTION
During the course of this lab assignment, it may be necessary to use various power tools and stepladders. It is imperative that a thorough understanding of the operating procedures and safety concerns associated with these power tools and stepladders be achieved before using them.

1. In the space provided here, sketch the basic floor plan of your wiring booth, and draw the wiring diagram that is needed to wire a lighting circuit that is controlled at one location, with the 120-volt source entering at the switch and two receptacles that are split, with half of each receptacle controlled by the switch. (Refer to Appendix I for an example floor plan.) The circuit should be laid out in the following manner:

   A. Switch should be located at the front of the left-hand panel and 4 ft (1.2 m) from the floor to the bottom of the box.

   B. Lighting outlet should be located overhead and in the center of the booth.

   C. One receptacle should be located at the front of the left-hand panel and 1 ft (305 mm) from the floor to the bottom of the box.

   D. The other receptacle should be located at the rear of the left-hand panel and 1 ft (305 mm) from the floor to the bottom of the box.

2. Calculate the minimum cubic-inch volume required for each of the boxes installed (using the appropriate 14 AWG NMB cable), and record your results here.

   Switch box   = _____

   Lighting outlet   = _____

   Receptacle outlet 1   = _____

   Receptacle outlet 2   = _____

3. Indicate in the spaces here the type and the minimum capacity (in.^{3}) of the boxes used in this lab (refer to *Table 314.16(A)* of the *NEC®*) (1 in.^{3} = 16.4 cm^3).

		Type of Box	Minimum Capacity (in.^{3})
Switch box	=	_____	_____
Lighting outlet	=	_____	_____
Receptacle outlet 1	=	_____	_____
Receptacle outlet 2	=	_____	_____

4. What are the acceptable colors of the grounding conductor? Where in the *NEC* is this stated?

_____

_____

5. What color is the grounding conductor that is being used to wire your electrical booth?

_____

_____

6. What is the minimum amount of cable assembly, including the sheath, that must extend into a nonmetallic box when using nonmetallic sheathed cable? Where in the *NEC* is this stated?

_____

_____

7. What does the *NEC* state about unused openings in nonmetallic boxes? Where in the *NEC* is this stated?

_____

_____

8. In the space provided here, draw the exact wiring that is involved in a lighting circuit that is controlled at one location, with the 120-volt source entering at the switch and two receptacles that are split, with half of each receptacle controlled by the switch. (Refer to Appendix II for an example wiring diagram.) Have the instructor verify that your diagram is drawn correctly and initial below.

CAUTION

It is important that the instructor check your wiring diagram before you begin to complete the electrical connections.

Instructor's initials: _____

9. Once the instructor has verified the validity of your wiring diagram, wire your assigned booth in accordance with the wiring diagram in step 8 (account for ½ in. [12.7 mm] thick gypsum board).

> **CAUTION**
> Do not work on circuits with power applied. Notify the instructor before energizing the circuit, and de-energize the power whenever possible.

10. Once you have completed the installation, have the instructor inspect and grade your wiring booth, using the following criteria:

Electrical integrity   _____   (Splices, terminations, and grounding were correctly done.)

NEC compliance   _____   (Local codes should be included.)

Operational   _____   (The circuit operates as defined.)

Appearance   _____   (Work shows a neat and professional manner.)

Safety factor   _____   (Student used proper safety practices and operated tools and equipment safely.)

> **CAUTION**
> Be sure to de-energize the electrical circuit before disassembling the wiring booth.

11. What question(s) do you still need answered concerning this lab assignment?

_____

_____

_____

12. What was the most interesting or important thing that you learned during this lab assignment?

_____

_____

_____

National Electrical Code® Drill Work

Select the correct answer for the following questions, and indicate the article or section in the *NEC* in which the answer can be found.

1. What is the maximum allowable ampacity of three single-insulated, 60°C rated, 8 AWG copper conductors in a raceway?

    A. 30 amperes

    B. 40 amperes   *NEC:* _____

    C. 50 amperes

    D. 55 amperes

2. A metal outlet box contains fourteen 14 AWG conductors. Which of the following is the minimum allowable box dimension?

    A.   4 in. × 1¼ in. square

    B.   4¹⁄₁₆ in. × 1¼ in. square   *NEC:* _____

    C.   4¹⁄₁₆ in. × 1½ in. square

    D.   4 in. × 2⅛ in. square

3. An equipment bonding jumper is permitted outside a raceway:

    A.   When not over 5 ft (1.5 m) long

    B.   When not over 6 ft (1.8 m) long   *NEC:* _____

    C.   When not over 8 ft (2.5 m) long

    D.   When not over 10 ft (3.0 m) long

4. A load is considered to be continuous if the maximum current is expected to continue for:

    A.   1 hour

    B.   2 hours   *NEC:* _____

    C.   2.5 hours

    D.   3 hours

5. What is the minimum number of 120-volt, 15-ampere, 2-wire lighting circuits required for a residence 85 ft (25.91 m) × 24 ft (7.32 m)?

    A.   2

    B.   3

    C.   4

    D.   5

6. In noncombustible walls or ceilings, the front edge of an outlet box may be set back of the finished surface not more than:

    A.   ⅛ in. (3.18 mm)

    B.   ¼ in. (6.35 mm)   *NEC:* _____

    C.   ½ in. (12.7 mm)

    D.   ¾ in. (19.05 mm)

Three-Way Switched Light Circuit with Feed at Switch

Name:	Date:	Grade:

Student Learning Outcomes

After the completing this lab, students should be able to:

- Perform the calculations involved in selecting the correctly sized boxes for the assigned task.

- Properly mount the boxes needed for a lighting circuit that is controlled at two locations, with the 120-volt source entering at the switch on the left-hand panel.

- Properly install the wiring, and connect the devices required for a lighting circuit that is controlled at two locations, with the 120-volt source entering at the switch on the left-hand panel.

Equipment and Supplies

Two 3-way switches, switch boxes

120-volt luminaire, lighting outlet

14/2 NMB cable

14/3 NMB cable

Cable staples

Procedures

CAUTION
Because portions of this lab involve working with 120 vac, your complete attention is required, and your every action must be well thought out and safety conscious.

POWER TOOL CAUTION
During the course of this lab assignment, it may be necessary to use various power tools and stepladders. It is imperative that a thorough understanding of the operating procedures and safety concerns associated with these power tools and stepladders be achieved before using them.

1. In the space provided here, sketch the basic floor plan of your wiring booth, and draw the wiring diagram that is needed to wire a lighting circuit that is controlled at two locations, with the 120-volt source entering at the switch on the left-hand panel. (Refer to Appendix I for an example floor plan.) The circuit should be laid out in the following manner:

   A. One switch should be located at the front of the left-hand panel and 4 ft (1.2 m) from the floor to the bottom of the box.

   B. The second switch should be located at the front of the right-hand panel and 4 ft (1.2 m) from the floor to the bottom of the box.

   C. The lighting outlet should be located overhead and in the center of the booth.

2. Calculate the minimum cubic-inch volume required for each of the boxes installed (using the appropriate 14 AWG NMB cable), and record your results here.

   1st switch box   =   _____

   2nd switch box   =   _____

   Lighting outlet   =   _____

3. Indicate in the spaces here the type and the minimum capacity (in.³) of the boxes used in this lab (refer to *Table 314.16(A)* of the *NEC®*) (1 in.³ = 16.4 cm³).

		Type of Box	**Minimum Capacity (in.³)**
1st switch box	=	_____	_____
2nd switch box	=	_____	_____
Lighting outlet	=	_____	_____

4. What are the acceptable colors of a 10 AWG grounded conductor? Where in the *NEC* is this stated?

_____

_____

5. What is (are) the color(s) of the grounded conductor(s) being used to wire your electrical booth?

_____

_____

6. When a grounded conductor in a residential setting is being used as an ungrounded conductor, what does the electrician need to do? Where in the *NEC* is this stated?

_____

_____

7. In the space provided here, draw the exact wiring that is involved in a lighting circuit that is controlled at two locations, with the 120-volt source entering at the switch on the left-hand panel. (Refer to Appendix II for an example wiring diagram.) Have the instructor verify that your diagram is drawn correctly and initial below.

CAUTION
It is important that the instructor check your wiring diagram before you begin to complete the electrical connections.

Instructor's initials: _____

8. Once the instructor has verified the validity of your wiring diagram, wire your assigned booth in accordance with the wiring diagram in step 7 (account for ½ in. [12.7 mm] thick gypsum board).

CAUTION
Do not work on circuits with power applied. Notify the instructor before energizing the circuit, and de-energize the power whenever possible.

9. Once you have completed the installation, have the instructor inspect and grade your wiring booth, using the following criteria:

Electrical integrity   _____   (Splices, terminations, and grounding were correctly done.)

NEC compliance   _____   (Local codes should be included.)

Operational   _____   (The circuit operates as defined.)

Appearance   _____   (Work shows a neat and professional manner.)

Safety factor   _____   (Student used proper safety practices and operated tools and equipment safely.)

CAUTION
Be sure to de-energize the electrical circuit before disassembling the wiring booth.

10. What did you find was the most challenging part of this lab assignment?

_____

_____

_____

11. Please describe any part of this lab that is still unclear to you.

_____

_____

_____

National Electrical Code® Drill Work

Select the correct answer for the following questions, and indicate the article or section in the *NEC* in which the answer can be found.

1. Which of the following statements about the mounting of electric equipment is (are) correct?
   I. Electric equipment shall be firmly secured to the surface on which it is mounted.
   II. Wooden plugs driven into holes in masonry, concrete, plaster, or similar materials shall not be used.
   A. I only
   B. II only   *NEC*: _____
   C. Both I and II
   D. Neither I nor II

2. A metal device box contains a duplex receptacle, cable clamps, and seven 12 AWG conductors. What is the minimum allowable size of the box?

   A. 20.25 in.³ (295.2 cm³)

   B. 22.5 in.³ (369 cm³)   *NEC*: _____

   C. 24.75 in.³ (406 cm³)

   D. 27.0 in.³ (443 cm³)

3. The maximum length of flexible cord identified for the purpose of connecting built-in dishwashers shall be:

   A. 18 in. (450 mm)

   B. 24 in. (600 mm)   *NEC*: _____

   C. 36 in. (900 mm)

   D. 48 in. (1.2 m)

4. What is the temperature rating for Type THHN insulation?

   A. 60°C

   B. 75°C   *NEC*: _____

   C. 90°C

   D. 105°C

5. Unless otherwise specifically permitted elsewhere in the *NEC*, which one of the following is the maximum overcurrent protection for three 14 AWG, Type THHN copper conductors installed in a raceway?

   A. 15 amperes

   B. 20 amperes   *NEC*: _____

   C. 25 amperes

   D. 30 amperes

Three-Way Switched Light Circuit with Feed at Light

Name:	Date:	Grade:

Student Learning Outcomes

After completing this lab, students should be able to:

- Perform the calculations involved in selecting the correctly sized boxes for the assigned task.

- Using both connection options, properly mount the boxes needed for a lighting circuit that is controlled at two locations, with the 120-volt source entering at the lighting outlet.

- Using both connection options, properly install the wiring, and connect the devices required for a lighting circuit that is controlled at two locations, with the 120-volt source entering at the lighting outlet.

Equipment and Supplies

Two 3-way switches, switch boxes

120-volt luminaire, lighting outlet

14/2 NMB cable

14/3 NMB cable

Cable staples

Procedures

CAUTION
Because portions of this lab involve working with 120 vac, your complete attention is required, and your every action must be well thought out and safety conscious.

POWER TOOL CAUTION
During the course of this lab assignment, it may be necessary to use various power tools and stepladders. It is imperative that a thorough understanding of the operating procedures and safety concerns associated with these power tools and stepladders be achieved before using them.

1. In the space provided here, sketch the basic floor plan of your wiring booth, and draw the wiring diagram that is needed to wire a lighting circuit that is controlled at two locations, with the 120-volt source entering at the lighting outlet, using one of the two connection options. (Refer to Appendix I for an example floor plan.) The circuit should be laid out in the following manner:

   A. One switch should be located at the front of the left-hand panel and 4 ft (1.2 m) from the floor to the bottom of the box.

   B. The second switch should be located at the front of the right-hand panel and 4 ft (1.2 m) from the floor to the bottom of the box.

   C. The luminaire should be located overhead and in the center of the booth.

2. Calculate the minimum cubic-inch volume required for each of the boxes installed (using the appropriate 14 AWG NMB cable), and record your results here.

   1st switch box   =   _____

   2nd switch box   =   _____

   Lighting outlet   =   _____

3. Indicate in the spaces here the type and the minimum capacity (in.^{3}) of the boxes used in this lab (refer to *Table 314.16(A)* of the *NEC®*) (1 in.^{3} = 16.4 cm^3).

		Type of Box	**Minimum Capacity (in.^{3})**
1st switch box	=	_____	_____
2nd switch box	=	_____	_____
Lighting outlet	=	_____	_____

4. What is the maximum distance that is allowed between cable straps or staples? Where in the *NEC* is this stated?

   _____

5. What is the minimum length of conductor required from when the conductor emerges from the cable or raceway or enters outlet to the end of the conductor? Where in the *NEC* is this stated?

_____

6. What is the minimum radius allowed along the inner edge of a bent cable? Where in the *NEC* is this stated?

_____

7. What is the maximum distance allowed between a nonmetallic box and the first cable strap or staple? Where in the *NEC* is this stated?

_____

8. In the space provided here, draw the exact wiring that is involved in a lighting circuit that is controlled by a 3-way switch, with the 120-volt source entering at the lighting outlet, using the connection method described in step 1. (Refer to Appendix II for an example wiring diagram.) Have the instructor verify that your diagram is drawn correctly and initial below.

CAUTION

It is important that the instructor check your wiring diagram before you begin to complete the electrical connections.

Instructor's initials: _____

9. Once the instructor has verified the validity of your wiring diagram, wire your assigned booth in accordance with the wiring diagram in step 8 (account for ½ in. [12.7 mm] thick gypsum board).

CAUTION

Do not work on circuits with power applied. Notify the instructor before energizing the circuit, and de-energize the power whenever possible.

10. Once you have completed the installation, have the instructor inspect and grade your wiring booth using the following criteria:

Electrical integrity _____ (Splices, terminations, and grounding were correctly done.)

NEC compliance _____ (Local codes should be included.)

Operational _____ (The circuit operates as defined.)

Appearance _____ (Work shows a neat and professional manner.)

Safety factor _____ (Student used proper safety practices and operated tools and equipment safely.)

CAUTION
Be sure to de-energize the electrical circuit before modifying the wiring booth.
---

11. In the space provided here, sketch the basic floor plan of your wiring booth, and using the other connection option, draw the wiring diagram that is needed to wire a lighting circuit that is controlled at two locations, with the 120-volt source entering at the lighting outlet. When you have completed the task, have the instructor verify that your diagram is drawn correctly and initial below.

A. One switch should be located at the front of the left-hand panel and 4 ft (1.2 m) from the floor to the bottom of the box.

B. The second switch should be located at the front of the right-hand panel and 4 ft (1.2 m) from the floor to the bottom of the box.

C. The luminaire should be located overhead and in the center of the booth.

CAUTION
It is important that the instructor check your wiring diagram before you begin to complete the electrical connections.

Instructor's initials: _____

12. Once the instructor has verified the validity of your wiring diagram, wire your assigned booth in accordance with the wiring diagram in step 11 (account for ½ in. [12.7 mm] thick gypsum board).

13. Once you have completed the installation, have the instructor inspect and grade your wiring booth, using the following criteria:

Electrical integrity   _____   (Splices, terminations, and grounding were correctly done.)

NEC compliance   _____   (Local codes should be included.)

Operational   _____   (The circuit operates as defined.)

Appearance   _____   (Work shows a neat and professional manner.)

Safety factor   _____   (Student used proper safety practices and operated tools and equipment safely.)

CAUTION

Be sure to de-energize the electrical circuit before disassembling the wiring booth.

---

14. What did you find was the most challenging part of this lab assignment?

_____

_____

_____

15. What is the most interesting or important thing that you learned during this lab assignment?

_____

_____

_____

National Electrical Code® Drill Work

Select the correct answer for the following questions, and indicate the article or section in the *NEC* in which the answer can be found.

1. What is the temperature rating for Type UF insulation with a single conductor?
   A. 60°C
   B. 75°C
   C. 90°C
   D. 105°C

   NEC: _____

2. Unless otherwise specifically permitted elsewhere in the *NEC*, which one of the following is the maximum overcurrent protection for three 12 AWG, Type THWN aluminum conductors installed in a raceway?

    A.   15 amperes

    B.   20 amperes   *NEC*: _____

    C.   25 amperes

    D.   30 amperes

3. When nailing a device outlet box, where the nail must pass through the interior of the enclosure, the nail must pass within what distance of the back or ends of the enclosure?

    A.   ⅛ in. (3.18 mm)

    B.   ¼ in. (6.35 mm)   *NEC*: _____

    C.   ½ in. (12.7 mm)

    D.   ¾ in. (19.05 mm)

4. The width of the working space in front of electric equipment shall be greater than either the width of the equipment or how many inches (millimeters)?

    A.   25 in. (635 mm)

    B.   30 in. (750 mm)   *NEC*: _____

    C.   35 in. (889 mm)

    D.   40 in. (1.02 m)

5. All of the following conductor insulations are rated for a dry and wet location, *except*:

    A.   TW

    B.   MI   *NEC*: _____

    C.   RHW-2

    D.   PFAH

6. In dwelling units, 20-ampere circuits are required in all of the following locations, *except*:

    A.   Bathrooms

    B.   Kitchens   *NEC*: _____

    C.   Laundry rooms

    D.   Outdoors

Four-Way Switched Light Circuit with Feed at Switch

Name:	Date:	Grade:

Student Learning Outcomes

After completing this lab, students should be able to:

- Perform the calculations involved in selecting the correctly sized boxes for the assigned task.

- Properly mount the boxes needed for a lighting circuit that is controlled at three locations, with the 120-volt source entering at the 3-way switch located on the right-hand panel.

- Properly install the wiring and connect the devices required for a lighting circuit that is controlled at three locations, with the 120-volt source entering at the 3-way switch located on the right-hand panel.

Equipment and Supplies

Two 3-way switches, switch boxes

One 4-way switch, switch box

120-volt luminaire, lighting outlet

14/2 NMB cable

14/3 NMB cable

Cable staples

Procedures

CAUTION
Because portions of this lab involve working with 120 vac, your complete attention is required, and your every action must be well thought out and safety conscious.

POWER TOOL CAUTION
During the course of this lab assignment, it may be necessary to use various power tools and stepladders. It is imperative that a thorough understanding of the operating procedures and safety concerns associated with these power tools and stepladders be achieved before using them.

1. In the space provided here, sketch the basic floor plan of your wiring booth, and draw the wiring diagram that must be used to wire a lighting circuit that is controlled at three locations, with the 120-volt source entering at the 3-way switch located on the right-hand panel. (Refer to Appendix I for an example floor plan.) The circuit should be laid out in the following manner:

   A.   One switch should be located at the front of the left-hand panel and 4 ft (1.2 m) from the floor to the bottom of the box.

   B.   The second switch should be located at the back of the left-hand panel and 4 ft (1.2 m) from the floor to the bottom of the box.

   C.   The third switch should be located at the back of the right-hand panel and 4 ft (1.2 m) from the floor to the bottom of the box.

   D.   The luminaire should be located overhead and in the center of the booth.

2. Calculate the minimum cubic-inch volume required for each of the boxes installed (using the appropriate 14 AWG NMB cable), and record your results here.

   1st switch box   =   _____

   2nd switch box   =   _____

   3rd switch box   =   _____

   Lighting outlet   =   _____

3. Indicate in the spaces here the type and the minimum capacity (in.^{3}) of the boxes used in this lab (refer to *Table 314.16(A)* of the *NEC®*) (1 in.^{3} = 16.4 cm^3).

		Type of Box	**Minimum Capacity (in.^{3})**
1st switch box	=	_____	_____
2nd switch box	=	_____	_____
3rd switch box	=	_____	_____
Lighting outlet	=	_____	_____

4. In the space provided on the next page, draw the exact wiring that is involved in a lighting circuit that is controlled at three locations, utilizing the connection method described in step 1. (Refer to Appendix II for an example wiring diagram.) Have the instructor verify that your diagram is drawn correctly and initial below.

CAUTION

It is important that the instructor check your wiring diagram before you begin to complete the electrical connections.

Instructor's initials: _____

5. Once the instructor has verified the validity of your wiring diagram, wire your assigned booth in accordance with the wiring diagram in step 4 (account for ½ in. [12.7 mm] thick gypsum board).

CAUTION

Do not work on circuits with power applied. Notify the instructor before energizing the circuit, and de-energize the power whenever possible.

6. Once you have completed the installation, have the instructor inspect and grade your wiring booth, using the following criteria:

Electrical integrity   _____   (Splices, terminations, and grounding were correctly done.)

NEC compliance   _____   (Local codes should be included.)

Operational   _____   (The circuit operates as defined.)

Appearance   _____   (Work shows a neat and professional manner.)

Safety factor   _____   (Student used proper safety practices and operated tools and equipment safely.)

CAUTION

Be sure to de-energize the electrical circuit before disassembling the wiring booth.

7. What did you find was the most challenging part of this lab assignment?

_____

_____

_____

8. What is the most interesting or important thing that you learned during this lab assignment?

_____

_____

_____

National Electrical Code® **Drill Work**

Select the correct answer for the following questions, and indicate the article or section in the *NEC* in which the answer can be found.

1. When repairing a noncombustible surface that was damaged, the maximum gap or open space at the edge of the box or fitting is equal to:
   A. ⅛ in. (3.18 mm)
   B. ¼ in. (6.35 mm)   *NEC:* _____
   C. ⅜ in. (9.53 mm)
   D. ½ in. (12.7 mm)

2. All of the following relate to SIS insulation, *except*:
   A. Its trade name is "Thermoset."
   B. It is rated at 90°C.   *NEC:* _____
   C. It is rated for both dry and wet locations.
   D. It can only be used for switchboard wiring.

3. What is the largest size conductor that can be installed in a raceway without having to be a stranded cable?
   A. 12 AWG
   B. 10 AWG   *NEC:* _____
   C. 8 AWG
   D. 6 AWG

4. A nonmetallic outlet box contains a 3-way switch, a duplex receptacle, and ten 14 AWG conductors. What is the minimum allowable size of the box?
   A. 20 in.³ (328 cm³)
   B. 24 in.³ (394 cm³)
   C. 26 in.³ (426 cm³)   *NEC:* _____
   D. 28 in.³ (459 cm³)

5. When using a wood brace as the primary enclosure support, what is the minimum allowable cross-section of the wood brace?
   A. ½ in. × 1 in. (12.7 mm × 25.4 mm)
   B. 1 in. × 1 in. (25.4 mm × 25.4 mm)   *NEC:* _____
   C. 1 in. × 2 in. (25.4 mm × 50.8 mm)
   D. 2 in. × 2 in. (50.8 mm × 50.8 mm)

6. How many 12 AWG THW conductors are permitted in a 4 in. × 1¼ in. octagonal metal box?
   A. 5
   B. 6   *NEC:* _____
   C. 8
   D. 9

Four-Way Switched Light Circuit with Feed at Light

Name:	Date:	Grade:

Student Learning Outcomes

After completing this lab, students should be able to:

- Perform the calculations involved in selecting the correctly sized boxes for the assigned task.

- Properly mount the boxes needed for a lighting circuit that is controlled at four locations, with the 120-volt source entering at the overhead lighting outlet.

- Properly install the wiring and connect the devices required for a lighting circuit that is controlled at four locations, with the 120-volt source entering at the overhead lighting outlet.

Equipment and Supplies

Two 3-way switches, switch boxes

Two 4-way switches, switch boxes

120-volt luminaire, lighting outlet

14/2 NMB cable

14/3 NMB cable

Cable staples

Procedures

CAUTION

Because portions of this lab involve working with 120 vac, your complete attention is required, and your every action must be well thought out and safety conscious.

POWER TOOL CAUTION

During the course of this lab assignment, it may be necessary to use various power tools and stepladders. It is imperative that a thorough understanding of the operating procedures and safety concerns associated with these power tools and stepladders be achieved before using them.

---

1. In the space provided here, sketch the basic floor plan of your wiring booth, and draw the wiring diagram that is needed to wire a lighting circuit that is controlled at four locations, with the 120-volt source entering at the overhead lighting outlet. (Refer to Appendix I for an example floor plan.) The circuit should be laid out in the following manner:

A. One switch should be located at the front of the left-hand panel and 14 ft (.2 m) from the floor to the bottom of the box.

B. The second switch should be located at the back of the left-hand panel and 4 ft (1.2 m) from the floor to the bottom of the box.

C. The third switch should be located at the front of the right-hand panel and 14 ft (.2 m) from the floor to the bottom of the box.

D. The fourth switch should be located at the back of the right-hand panel and 4 ft (1.2 m) from the floor to the bottom of the box.

E. The luminaire should be located overhead and in the center of the booth.

2. Calculate the minimum cubic-inch volume required for each of the boxes installed (using the appropriate 14 AWG NMB cable), and record your results here.

1st switch box   =   _____

2nd switch box   =   _____

3rd switch box   =   _____

4th switch box   =   _____

Lighting outlet   =   _____

3. Indicate in the spaces here the type and the minimum capacity (in.^{3}) of the boxes used in this lab (refer to *Table 314.16(A)* of the *NEC®*) (1 in.^{3} = 16.4 cm^3).

		Type of Box	Minimum Capacity (in.^{3})
1st switch box	=	_____	_____
2nd switch box	=	_____	_____
3rd switch box	=	_____	_____
4th switch box	=	_____	_____
Lighting outlet	=	_____	_____

4. In the space provided here, draw the exact wiring that is involved in a lighting circuit that is controlled at three locations, with the 120-volt source entering at the overhead lighting outlet, using the connection method described in step 1. (Refer to Appendix II for an example wiring diagram.) Have the instructor verify that your diagram is drawn correctly and initial below.

CAUTION
It is important that the instructor check your wiring diagram before you begin to complete the electrical connections.

Instructor's initials: _____

5. Once the instructor has verified the validity of your wiring diagram, wire your assigned booth in accordance with the wiring diagram in step 4 (account for ½ in. [12.7 mm] thick gypsum board).

CAUTION
Do not work on circuits with power applied. Notify the instructor before energizing the circuit, and de-energize the power whenever possible.

6. Once you have completed the installation, have the instructor inspect and grade your wiring booth, using the following criteria:

Electrical integrity   _____   (Splices, terminations, and grounding were correctly done.)

NEC compliance   _____   (Local codes should be included.)

Operational   _____   (The circuit operates as defined.)

Appearance   _____   (Work shows a neat and professional manner.)

Safety factor   _____   (Student used proper safety practices and operated tools and equipment safely.)

CAUTION
Be sure to de-energize the electrical circuit before disassembling the wiring booth.

7. What did you find was the most challenging part of this lab assignment?

_____

_____

_____

8. What is the most interesting or important thing that you learned during this lab assignment?

_____

_____

_____

National Electrical Code® **Drill Work**

Select the correct answer for the following questions, and indicate the article or section in the *NEC* in which the answer can be found.

1. What is characterized by the use of the terms "shall be permitted" or "shall not be permitted"?
   A. Mandatory rules
   B. Permissive rules   *NEC*: _____
   C. Explanatory material
   D. Optional material

2. No outlet box shall have an internal depth less than:
   A. ⅛ in. (3.18 mm)
   B. ¼ in. (6.35 mm)   *NEC*: _____
   C. ½ in. (12.7 mm)
   D. ¹⁵⁄₁₆ in. (23.9 mm)

3. How many 16 AWG THHN insulated copper conductors are permitted in a $3 \times 2 \times 2$ metal device box?
   A. 4
   B. 5   *NEC*: _____
   C. 6
   D. 7

4. The *NEC* covers the following installations, *except*:
   A. Installations of optical fiber cable
   B. Installations in buildings used by the electric utility   *NEC*: _____
   C. Installations underground in mines
   D. Installations of electric conductors within private buildings

5. What is the minimum size copper conductor that can be used for an application when the conductor has a voltage rating between 2001 and 5000 volts?

    A.   8 AWG

    B.   10 AWG   *NEC*: _____

    C.   12 AWG

    D.   14 AWG

6. Partially protected locations under canopies and roofed open porches are referred to as:

    A.   Wet locations

    B.   Dry locations   *NEC*: _____

    C.   Damp locations

    D.   None of the above

Ground-Fault Circuit Interrupt Wiring

Name:	Date:	Grade:

Student Learning Outcomes

After completing this lab, students should be able to:

- Perform the calculations involved in selecting the correctly sized boxes for the assigned task.

- Properly mount the boxes needed for an electrical circuit that contains a light circuit that is controlled at one location, with the 120-volt source entering at the switch and a continuously energized GFCI receptacle circuit containing two duplex receptacles.

- Properly install the wiring and connect the devices required for an electrical circuit that contains a light circuit that is controlled at one location, with the 120-volt source entering at the switch and a continuously energized GFCI receptacle circuit containing two duplex receptacles.

Equipment and Supplies

Single-pole switch, switch box

GFCI duplex receptacle, receptacle outlet

Duplex receptacle, receptacle outlet

120-volt luminaire, lighting outlet

14/2 NMB cable

14/3 NMB cable

Cable staples

Procedures

CAUTION
Because portions of this lab involve working with 120 vac, your complete attention is required, and your every action must be well thought out and safety conscious.

POWER TOOL CAUTION
During the course of this lab assignment, it may be necessary to use various power tools and stepladders. It is imperative that a thorough understanding of the operating procedures and safety concerns associated with these power tools and stepladders be achieved before using them.

1. In the space provided here, sketch the basic floor plan of your wiring booth, and draw the wiring diagram that is needed to wire an electrical circuit that is controlled at one location, with the 120-volt source entering at the switch and a continuously energized GFCI receptacle circuit containing two duplex receptacles. (Refer to Appendix I for an example floor plan.) The circuit should be laid out in the following manner:

   A. The switch should be located at the front of the left-hand panel and 4 ft (1.2 m) from the floor to the bottom of the box.

   B. The luminaire should be located overhead and in the center of the booth.

   C. One duplex receptacle should be located directly below the light switch and 1 ft (305 mm) from the floor to the bottom of the box.

   D. The second duplex receptacle should be located at the rear of the left-hand panel and 1 ft (305 mm) from the floor to the bottom of the box.

2. Calculate the minimum cubic-inch volume required for each of the boxes installed (using the appropriate 14 AWG NMB cable), and record your results here.

   Switch box   =   _____

   1st receptacle outlet   =   _____

   2nd receptacle outlet   =   _____

   Lighting outlet   =   _____

3. Indicate in the spaces here the type and the minimum capacity (in.³) of the boxes used in this lab (refer to *Table 314.16(A)* of the *NEC®*) (1 in.³ = 16.4 cm³).

		Type of Box	**Minimum Capacity (in.³)**
Switch box	=	_____	_____
1st receptacle outlet	=	_____	_____
2nd receptacle outlet	=	_____	_____
Lighting outlet	=	_____	_____

4. In the space provided here, and using the connection method described in step 1, draw the exact wiring that is involved in an electrical circuit that contains a light circuit that is controlled at one location, with the 120-volt source entering at the switch and a continuously energized GFCI receptacle circuit containing two duplex receptacles. (Refer to Appendix II for an example wiring diagram.) Have the instructor verify that your diagram is drawn correctly and initial below.

> **CAUTION**
> It is important that the instructor check your wiring diagram before you begin to complete the electrical connections.

Instructor's initials: _____

5. Once the instructor has verified the validity of your wiring diagram, wire your assigned booth in accordance with the wiring diagram in step 4 (account for ½ in. [12.7 mm] thick gypsum board).

> **CAUTION**
> Do not work on circuits with power applied. Notify the instructor before energizing the circuit, and de-energize the power whenever possible.

6. Once you have completed the installation, have the instructor inspect and grade your wiring booth, using the following criteria:

Electrical integrity   _____   (Splices, terminations, and grounding were correctly done.)

NEC compliance   _____   (Local codes should be included.)

Operational   _____   (The circuit operates as defined.)

Appearance   _____   (Work shows a neat and professional manner.)

Safety factor   _____   (Student used proper safety practices and operated tools and equipment safely.)

> **CAUTION**
> Be sure to de-energize the electrical circuit before disassembling the wiring booth.

7. What did you find was the most challenging part of this lab assignment?

_____

_____

_____

8. What is the most interesting or important thing that you learned during this lab assignment?

_____

_____

_____

National Electrical Code® Drill Work

Select the correct answer for the following questions, and indicate the article or section in the *NEC* in which the answer can be found.

1. Boxes that do not enclose devices shall have an internal depth of no less than:
   A. ⅛ in. (3.18 mm)
   B. ¼ in. (6.35 mm)   *NEC:* _____
   C. ½ in. (12.7 mm)
   D. ¹⁵⁄₁₆ in. (23.9 mm)

2. What is defined as a point on a wiring system when current is taken to supply equipment that uses electricity?
   A. Outlet
   B. Device   *NEC:* _____
   C. Box
   D. Receptacle

3. What is characterized by the use of the terms *shall* and *shall not*?
   A. Mandatory rules
   B. Permissive rules   *NEC:* _____
   C. Explanatory material
   D. Optional material

4. What is the minimum number of 120-volt, 15-ampere lighting branch circuits required for a one-story residence that is 87 ft × 42 ft (26.5 m × 12.8 m) and *includes* a one-floor garage that measures 18 ft × 22 ft (5.5 m × 6.7 m)?
   A. 5
   B. 6
   C. 7
   D. 8

5. If an outlet box is to contain six 10 AWG TW insulated conductors and four 10 AWG XHHW-2 insulated conductors, which metal outlet would be required?

   A.   4 in. × $1\frac{1}{2}$ in. round

   B.   4 in. × $1\frac{1}{2}$ in. square   *NEC:* _____

   C. $4^{11}/_{16}$ in. × $1\frac{1}{2}$ in. square

   D.   4 in. × $2\frac{1}{2}$ in. round

6. A location not normally subject to dampness or wetness is referred to as a:

   A.   Wet location

   B.   Dry location   *NEC:* _____

   C.   Damp location

Combination Circuit Lab #1

Name:	Date:	Grade:

Student Learning Outcomes

After completing this lab, students should be able to:

- Perform the calculations involved in selecting the correctly sized boxes for the assigned task.

- Properly mount the boxes needed for the following electrical circuits:

    - An overhead light circuit that is controlled at two locations.

    - A split duplex-receptacle circuit with two duplex receptacles that are controlled at one location.

    A 2-gang box is used to house the receptacle circuit switch and one of the switches of the light circuit.

- Properly install the wiring and connect the devices required by the electrical circuits that are listed above.

Equipment and Supplies

Single-pole switch, switch box

Two 3-way switches, switch boxes

Two duplex receptacles, receptacle outlets

120-volt luminaire, lighting outlet

14/2 NMB cable

14/3 NMB cable

Cable staples

Procedures

CAUTION

Because portions of this lab involve working with 120 vac, your complete attention is required, and your every action must be well thought out and safety conscious.

POWER TOOL CAUTION

During the course of this lab assignment, it may be necessary to use various power tools and stepladders. It is imperative that a thorough understanding of the operating procedures and safety concerns associated with these power tools and stepladders be achieved before using them.

1. In the space provided here, sketch the basic floor plan of your wiring booth, and draw the wiring diagram required to install the electrical circuits described here. The feed to the circuits should enter at the switches located at the front of the left-hand panel. (Refer to Appendix I for an example floor plan.) The circuit should be laid out in the following manner:

A. An overhead light circuit that is controlled at two locations and is to be laid out in the following manner:

   1. One switch should be located at the front of the left-hand panel and 4 ft (1.2 m) from the floor to the bottom of the box, and the 120-volt source is to enter at this switch.

   2. The other switch should be located at the front of the right-hand panel and 4 ft (1.2 m) from the floor to the bottom of the box.

   3. The overhead lighting outlet should be located overhead and in the center of the booth.

B. A split duplex-receptacle circuit with two duplex receptacles that are controlled at one location.

   1. The switch should be located at the front of the left-hand panel and 4 ft (1.2 m) from the floor to the bottom of the box and the 120-volt source is to enter at this switch.

   2. One duplex receptacle should be located directly below the light switch and 1 ft (305 mm) from the floor to the bottom of the box.

   3. The other duplex receptacle should be located at the rear of the left-hand panel of the booth and 1 ft (305 mm) from the floor to the bottom of the box.

2. Calculate the minimum cubic-inch volume required for each of the boxes installed (using the appropriate 14 AWG NMB cable), and record your results here.

Overhead Light Circuit

1st switch box   =   _____

2nd switch box   =   _____

Lighting outlet   =   _____

Receptacle Outlet

Switch box   =   _____

1st receptacle outlet   =   _____

2nd receptacle outlet   =   _____

3. Indicate in the spaces here the type and the minimum capacity (in.³) of the boxes used in this lab (refer to *Table 314.16(A)* of the *NEC®*) (1 in.³ = 16.4 cm³).

	Type of Box	**Minimum Capacity (in.³)**
Overhead Light Circuit		
1st switch box =	_____	_____
2nd switch box =	_____	_____
Lighting outlet =	_____	_____
Receptacle Outlet		
Switch box =	_____	_____
1st receptacle outlet =	_____	_____
2nd receptacle outlet =	_____	_____

4. In the space provided here, and using the connection methods described in step 1, draw the exact wiring that is involved in the electrical circuits specified in step 1. (Refer to Appendix II for an example wiring diagram.) Once this task has been completed, have the instructor verify that your diagram is drawn correctly and initial below.

CAUTION
It is important that the instructor check your wiring diagram before you begin to complete the electrical connections.

Instructor's initials: _____

5. Once the instructor has verified the validity of your wiring diagram, wire your assigned booth in accordance with the wiring diagram in step 4 (account for ½ in. [12.7 mm] thick gypsum board).

> **CAUTION**
> Do not work on circuits with power applied. Notify the instructor before energizing the circuit, and de-energize the power whenever possible.

6. Once you have completed the installation, have the instructor inspect and grade your wiring booth, using the following criteria:

Electrical integrity	_____	(Splices, terminations, and grounding were correctly done.)
NEC compliance	_____	(Local codes should be included.)
Operational	_____	(The circuit operates as defined.)
Appearance	_____	(Work shows a neat and professional manner.)
Safety factor	_____	(Student used proper safety practices and operated tools and equipment safely.)

> **CAUTION**
> Be sure to de-energize the electrical circuit before disassembling the wiring booth.

7. What did you find was the most challenging part of this lab assignment?

_____

_____

_____

8. What is the most interesting or important thing that you learned during this lab assignment?

_____

_____

_____

National Electrical Code® Drill Work

Select the correct answer for the following questions listed, and indicate the article or section in the NEC in which the answer can be found.

1. True or false? In electrical equipment rooms, the illumination shall not be controlled by automatic means only.
   A. True
   B. False   NEC: _____

2. What is the ampacity of a 10 AWG RHW conductor if it is run in a walk-in freezer that has an ambient temperature of 22°C?

   A.   26.25 amperes

   B.   33.33 amperes   *NEC:* _____

   C.   36.75 amperes

   D.   52.51 amperes

3. All of the following articles have been changed since the 2011 *NEC, except*:

   A.   *426.50*

   B.   *610.8*   *NEC:* _____

   C.   *398.17*

   D.   *310.104*

4. A 120/240-volt, 3-wire service drop passing over a pedestrian sidewalk shall have a minimum clearance from final grade of how many feet?

   A.   10 ft (3.0 m)

   B.   12 ft (3.6 m)   *NEC:* _____

   C.   15 ft (4.5 m)

   D.   18 ft (4.8 m)

5. The minimum number of inches (millimeters) between the top surface of a finished grade and the top of a direct burial cable that is in a trench that is beneath a 3 in. (75 mm) thick concrete slab is:

   A.   6 in. (150 mm)

   B.   12 in. (300 mm)   *NEC:* _____

   C.   18 in. (450 mm)

   D.   24 in. (600 mm)

6. Determine the minimum cubic centimeters (cubic inches) required for three 12 AWG TW conductors passing through a box, five 14 AWG THHN conductors terminating in the box, two 10 AWG TW conductors terminating on a receptacle, and one 10 AWG equipment bonding jumper from the receptacle to the box.

   A.   21.0 in.^{3} (344 cm^3)

   B.   24.0 in.^{3} (395 cm^3)   *NEC:* _____

   C.   25.5 in.^{3} (418 cm^3)

   D.   29.25 in.^{3} (480 cm^3)

Combination Circuit Lab #2

Name:	Date:	Grade:

Student Learning Outcomes

After completing this lab, students should be able to:

- Perform the calculations involved in selecting the correctly sized boxes for the assigned task.
- Properly mount the boxes needed for the following electrical circuits:
  - An overhead light circuit that is controlled at three locations, with the feed at the lighting outlet.
  - A wall-mounted light circuit that is controlled in one location.
  - A GFCI-protected duplex-receptacle circuit.
- Properly install the wiring and connect the devices required by the electrical circuits that are listed above.

Equipment and Supplies

Single-pole switch, switch box

Two 3-way switches, switch boxes

Four-way switch, switch box

Duplex receptacle, receptacle outlet

GFCI duplex receptacle, receptacle outlet

Two 120-volt luminaires, lighting outlets

14/2 NMB cable

14/3 NMB cable

Cable staples

Procedures

<table>
<tr><td>CAUTION
Because portions of this lab involve working with 120 vac, your complete attention is required, and your every action must be well thought out and safety conscious.</td></tr>
</table>

<table>
<tr><td>POWER TOOL CAUTION
During the course of this lab assignment, it may be necessary to use various power tools and stepladders. It is imperative that a thorough understanding of the operating procedures and safety concerns associated with these power tools and stepladders be achieved before using them.</td></tr>
</table>

73

1. In the space provided here, sketch the basic floor plan of your wiring booth, and draw the wiring diagram required to install the electrical circuits described here. (Refer to Appendix I for an example floor plan.) The circuit should be laid out in the following manner:

   A. An overhead light circuit that is controlled at three locations and is to be laid out as described here:

      1. One switch should be located at the front of the left-hand panel and 4 ft (1.2 m) from the floor to the bottom of the ganged box.

      2. The second switch should be located at the rear of the left-hand panel and 4 ft (1.2 m) from the floor to the bottom of the box.

      3. The third switch should be located at the rear of the right-hand panel and 4 ft (1.2 m) from the floor to the bottom of the box.

      4. The overhead luminaire should be located overhead and in the center of the booth, and the 120-volt source should enter the circuit at this light receptacle.

   B. A wall-mounted light circuit that is controlled at one location and is to be laid out as described here:

      1. The switch should be located in a ganged box at the front of the left-hand panel and 4 ft (1.2 m) from the floor to the bottom of the box; the 120-volt source should feed this circuit at this point.

      2. The wall-mounted luminaire should be mounted on the front of the left-hand panel and 6 ft (1.8 m) from the floor to the bottom of the box.

   C. A GFCI duplex-receptacle circuit that is to be laid out as described here:

      1. The first receptacle should be located at the front of the left-hand panel and 1 ft (305 mm) from the floor to the bottom of the box, and the 120-volt source should feed this circuit at this point.

      2. The second receptacle should be located at the rear of the left-hand panel and 1 ft (305 mm) from the floor to the bottom of the box.

2. Calculate the minimum cubic-inch volume required for each of the boxes installed (using the appropriate 14 AWG NMB cable), and record your results here.

Overhead Light Circuit

1st switch box	=	_____
2nd switch box	=	_____
3rd switch box	=	_____
Lighting outlet	=	_____

Wall-Mounted Light Circuit

Switch box	=	_____
Lighting outlet	=	_____

GFCI Receptacle Circuit

1st receptacle outlet	=	_____
2nd receptacle outlet	=	_____

3. Indicate in the spaces here the type and the minimum capacity (in.^{3}) of the boxes used in this lab (refer to *Table 314.16(A)* of the *NEC®*) (1 in.^{3} = 16.4 cm^3).

	Type of Box	Minimum Capacity (in.^{3})

Overhead Light Circuit

1st switch box	=	_____	_____
2nd switch box	=	_____	_____
3rd switch box	=	_____	_____
Lighting outlet	=	_____	_____

Wall-Mounted Light Circuit

Switch box	=	_____	_____
Lighting outlet	=	_____	_____

GFCI Receptacle Circuit

1st receptacle outlet	=	_____	_____
2nd receptacle outlet	=	_____	_____

4. In the space provided, and using the connection methods described in step 1, draw the exact wiring that is involved in the electrical circuits specified in step 1. (Refer to Appendix II for an example wiring diagram.) Have the instructor verify that your diagram is drawn correctly and initial below.

CAUTION

It is important that the instructor check your wiring diagram before you begin to complete the electrical connections.

Instructor's initials: _____

5. Once the instructor has verified the validity of your wiring diagram, wire your assigned booth in accordance with the wiring diagram in step 4 (account for ½ in. 12.7 mm] thick gypsum board).

CAUTION

Do not work on circuits with power applied. Notify the instructor before energizing the circuit, and de-energize the power whenever possible.

---

6. Once you have completed the installation, have the instructor inspect and grade your wiring booth, using the following criteria:

Electrical integrity	_____	(Splices, terminations, and grounding were correctly done.)
NEC compliance	_____	(Local codes should be included.)
Operational	_____	(The circuit operates as defined.)
Appearance	_____	(Work shows a neat and professional manner.)
Safety factor	_____	(Student used proper safety practices and operated tools and equipment safely.)

CAUTION

Be sure to de-energize the electrical circuit before disassembling the wiring booth.

---

7. What did you find was the most challenging part of this lab assignment?

_____

_____

_____

8. What is the most interesting or important thing that you learned during this lab assignment?

_____

_____

_____

National Electrical Code® Drill Work

Select the correct answer for the following questions, and indicate the article or section in the *NEC* in which the answer can be found.

1. Which of the following duty operations is defined as "a substantially constant load for a short and definitely specified time"?
   A. Continuous duty
   B. Periodic duty   *NEC:* _____
   C. Short-time duty
   D. Varying duty

2. The NM type nonmetallic sheathed cable can be installed in all of the following, *except*:
   A. One-family dwellings
   B. Identified cable trays   *NEC:* _____
   C. Two-family dwellings
   D. Where exposed to corrosive fumes

3. What is the minimum size for service-entrance conductors for a service exceeding 600 volts in a multiconductor cable?
   A. 4 AWG
   B. 6 AWG   *NEC:* _____
   C. 8 AWG
   D. 10 AWG

4. When the concealed knob-and-tube wiring method is used, the conductors must be supported within how many inches (millimeters) of each tap or splice?
   A. 4 in. (100 mm)
   B. 6 in. (150 mm)   *NEC:* _____
   C. 8 in. (200 mm)
   D. 12 in. (300 mm)

5. Which of the following articles has been changed since the 2011 *NEC*?
    A.   *700.8*
    B.   *392.46*
    C.   *240.12*
    D.   *400.8*

6. A lighting track is not allowed in the following locations, *except*:
    A.   6 ft (1.8 m) above the finished floor
    B.   In a damp location   *NEC*: _____
    C.   In a storage battery room
    D.   Where subject to corrosive vapors

Combination Circuit Lab #3

Name:	Date:	Grade:

Student Learning Outcomes

After completing this lab, students should be able to:

- Perform the calculations involved in selecting the correctly sized boxes for the assigned electrical circuits.

- Properly mount the boxes needed for the following electrical circuits:
  - An overhead light circuit that is controlled at four locations.
  - A wall-mounted light circuit that is controlled at one location.
  - A split duplex-receptacle circuit with two duplex receptacles that are controlled at one location.

- Properly install the wiring and connect the devices required by the electrical circuits that are listed above.

Equipment and Supplies

Two single-pole switches, switch boxes

Two 3-way switches, switch boxes

Two 4-way switches, switch boxes

Two duplex receptacles, receptacle outlets

Two 120-volt luminaires, lighting outlets

14/2 NMB cable

14/3 NMB cable

Cable staples

Procedures

CAUTION

Because portions of this lab involve working with 120 vac, your complete attention is required, and your every action must be well thought out and safety conscious.

POWER TOOL CAUTION

During the course of this lab assignment, it may be necessary to use various power tools and stepladders. It is imperative that a thorough understanding of the operating procedures and safety concerns associated with these power tools and stepladders be achieved before using them.

1. In the space provided here, sketch the basic floor plan of your wiring booth, and draw the wiring diagram required to install the electrical circuits described here. (Refer to Appendix I for an example floor plan.) The circuit should be laid out in the following manner:

A. An overhead light circuit that is controlled at four locations and is to be laid out in the following manner:

   1. Two switches are located on the left-hand panel, one at the front and one at the back of the panel, and 4 ft (1.2 m) from the floor to the bottom of the box; the 120-volt source should enter this circuit at the switch at the front of the left-hand panel.

   2. The other two switches should be located on the right-hand panel, one at the front and one at the back of the panel and 4 ft (1.2 m) from the floor to the bottom of the box.

B. A wall-mounted light circuit that is controlled at one location and is to be laid out in the following manner:

   1. The luminaire should be located on the outside edge of the left-hand panel and 5 ft (1.5 m) from the ground.

   2. The switch should be located at the front of the left-hand panel, and the 120-volt source should enter this circuit at this switch.

C. A split duplex-receptacle circuit with two duplex receptacles that are controlled at one location, and which is to be laid out in the following manner:

   1. The switch should be located at the front of the left-hand panel and 4 ft (1.2 m) from the floor to the bottom of the box, and the 120-volt source should enter this circuit at this switch.

   2. One duplex receptacle should be located directly below the light switch and 1 ft (300 mm) from the floor to the bottom of the box.

   3. The other duplex receptacle should be located at the rear of the left-hand panel and 1 ft (300 mm) from the floor to the bottom of the box.

A 3-gang box is used to house the switches of the three circuits that are located at the front of the left-hand panel.

2. Calculate the minimum cubic-inch volume required for each of the boxes installed (using the appropriate 14 AWG NMB cable), and record your results here.

Overhead Light Circuit

Switch box 1   =   _____

Switch box 2   =   _____

Switch box 3   =   _____

Switch box 4   =   _____

Lighting outlet   =   _____

Wall-Mounted Light Circuit

Switch box   =   _____

Lighting outlet   =   _____

Receptacle Circuit

Switch box   =   _____

1st receptacle outlet =   _____

2nd receptacle outlet =   _____

3. Indicate in the spaces here the type and the minimum capacity (in.³) of the boxes used in this lab (refer to *Table 314.16(A)* of the *NEC®*) (1 in.³ = 16.4 cm³).

	Type of Box	Minimum Capacity (in.³)
Overhead Light Circuit		
Switch box 1 =	_____	_____
Switch box 2 =	_____	_____
Switch box 3 =	_____	_____
Switch box 4 =	_____	_____
Lighting outlet =	_____	_____

**Wall-Mounted Light
Circuit**

Switch box   = _____   _____

Lighting outlet   = _____   _____

Receptacle Circuit

Switch box   = _____   _____

1st receptacle outlet   = _____   _____

2nd receptacle outlet   = _____   _____

4. In the space provided here, and using the connection methods described in step 1, draw the exact wiring that is involved in the electrical circuits specified in step 1. (Refer to Appendix II for an example wiring diagram.) Have the instructor verify that your diagram is drawn correctly and initial below.

CAUTION

It is important that the instructor check your wiring diagram before you begin to complete the electrical connections.

Instructor's initials: _____

5. Once the instructor has verified the validity of your wiring diagram, wire your assigned booth in accordance with the wiring diagram in step 4 (account for ½ in. [12.7 mm] thick gypsum board).

CAUTION

Do not work on circuits with power applied. Notify the instructor before energizing the circuit, and de-energize the power whenever possible.

---

6. Once you have completed the installation, have the instructor inspect grade and your wiring booth, using the following criteria:

Electrical integrity   _____   (Splices, terminations, and grounding were correctly done.)

NEC compliance   _____   (Local codes should be included.)

Operational   _____   (The circuit operates as defined.)

Appearance   _____   (Work shows a neat and professional manner.)

Safety factor   _____   (Student used proper safety practices and operated tools and equipment safely.)

CAUTION
Be sure to de-energize the electrical circuit before disassembling the wiring booth.

7. What did you find was the most challenging part of this lab assignment?

_____

_____

_____

8. What is the most interesting or important thing that you learned during this lab assignment?

_____

_____

_____

National Electrical Code® Drill Work

Select the correct answer for the following questions, and indicate the article or section in the NEC in which the answer can be found.

1. Which of the following statements is not applicable to TW conductor insulation?
   A. It is suitable for dry and wet locations.
   B. It is flame-retardant.   NEC: _____
   C. It is moisture-resistant.
   D. It is suitable for damp and wet locations.

2. What is the minimum size service-entrance conductor for services exceeding 600 volts and that are *not* in a multiconductor cable?
   A. 4 AWG
   B. 6 AWG   NEC: _____
   C. 8 AWG
   D. 10 AWG

3. An 8-foot feeder tap of an ungrounded conductor must have overcurrent protection at the tap if:

   A.   The tap conductors extend beyond the panelboard.

   B.   The tap conductors are in a raceway.   *NEC*: _____

   C.   The tap conductors do not extend beyond the switchboard.

   D.   The ampacity of the tap conductor is more than combined computed loads of the supplied circuits.

4. What is the minimum number of 120-volt, 15-ampere lighting circuits required for a one-story residence with an outside dimension of 42 ft × 75 ft (12.8 m × 22.9 m) that includes a 3.0 m × 3.0 m (10 ft × 10 ft) front porch?

   A.   4

   B.   5

   C.   6

   D.   7

5. When electrical cabinets are installed in concrete walls, the front edge of the cabinet cannot be set back from the walls' finished surface by more than how many inches (millimeters)?

   A.   ⅛ in. (3.18 mm)

   B.   ¼ in. (6.35 mm)   *NEC*: _____

   C.   ½ in. (12.7 mm)

   D.   ¾ in. (19.05 mm)

6. How does the NEC define the phrase in sight from?

_____

_____

_____

Combination Circuit Lab #4

Name:	Date:	Grade:

Student Learning Outcomes

After completing this lab, students should be able to:

- Perform the calculations involved in selecting the correctly sized boxes for the assigned electrical circuits.

- Properly mount the boxes needed for the following electrical circuits:

  - An overhead light circuit that is controlled at two locations.
  - A duplex-receptacle circuit with three duplex receptacles in which only two of the receptacles are GFCI protected.
  - A split duplex-receptacle circuit with three duplex receptacles that are controlled at one location.

- Properly install the wiring and connect the devices required by the electrical circuits that are listed above.

Equipment and Supplies

Four single-pole switches, switch boxes

GFCI duplex receptacle, receptacle outlet

Five duplex receptacles, receptacle outlets

120-volt luminaire, lighting outlet

14/2 NMB cable

14/3 NMB cable

Cable staples

Procedures

> **CAUTION**
> Because portions of this lab involve working with 120 vac, your complete attention is required, and your every action must be well thought out and safety conscious.

> **POWER TOOL CAUTION**
> During the course of this lab assignment, it may be necessary to use various power tools and stepladders. It is imperative that a thorough understanding of the operating procedures and safety concerns associated with these power tools and stepladders be achieved before using them.

1. In the space provided here, sketch the basic floor plan of your wiring booth, and draw the wiring diagram that is required to wire the electrical circuits that are described here. (Refer to Appendix I for an example floor plan.) The circuit should be laid out in the following manner:

A. An overhead light circuit that is controlled at two locations, and in which the 120-volt source is to enter at the switch on the left-hand panel:

   1. One switch should be located at the front of the left-hand panel and 4 ft (1.2 m) from the floor to the bottom of the box.

   2. The second switch should be located at the front of the right-hand panel and 4 ft (1.2 m) from the floor to the bottom of the box.

   3. The luminaire should be located overhead and in the center of the booth.

B. A duplex-receptacle circuit with three duplex receptacles in which only two of the receptacles are GFCI protected.

   1. The non-GFCI-protected receptacle should be located at the back of the left-hand panel, and the 120-volt source is to enter at this receptacle.

   2. The two GFCI-protected receptacles should be located at the front and back of the right-hand panel.

All of the receptacles should be mounted 1 ft (305 mm) from the floor of the booth.

C. A split duplex-receptacle circuit with three duplex receptacles that are controlled at one location.

   1. The switch is to be located on the front of the left-hand panel, and the 120-volt source is to enter at this switch.

   2. The first receptacle should be placed at the front of the left-hand panel, directly below the switch.

   3. The second receptacle should be placed at the back of the left-hand panel.

   4. The third receptacle should be placed at the front of the right-hand panel.

All of the receptacles should be mounted 1 ft (305 mm) from the floor of the booth.

2. Calculate the minimum cubic-inch volume required for each of the boxes installed (using the appropriate 14 AWG NMB cable), and record your results here.

Overhead Light Circuit

1st switch box   =   _____

2nd switch box   =   _____

Lighting outlet   =   _____

GFCI Receptacle Circuit

1st receptacle outlet   =   _____

2nd receptacle outlet   =   _____

3rd receptacle outlet   =   _____

Receptacle Circuit

Switch box   =   _____

1st receptacle outlet   =   _____

2nd receptacle outlet   =   _____

3rd receptacle outlet   =   _____

3. In the space provided here, and using the electrical circuit described in step 1, draw the exact wiring that is involved in physically connecting the electrical circuit. (Refer to Appendix II for an example wiring diagram.) Have the instructor verify that your diagram is drawn correctly and initial below.

CAUTION

It is important that the instructor check your wiring diagram before you begin to complete the electrical connections.

Instructor's initials: _____

4. Once the instructor has verified the validity of your wiring diagram, wire your assigned booth in accordance with the wiring diagram in step 3 (account for ½ in. [12.7 mm] thick gypsum board).

CAUTION

Do not work on circuits with power applied. Notify the instructor before energizing the circuit, and de-energize the power whenever possible.

---

5. Once you have completed the installation, have the instructor inspect and grade your wiring booth, using the following criteria:

Electrical integrity	_____	(Splices, terminations, and grounding were correctly done.)
NEC® compliance	_____	(Local codes should be included.)
Operational	_____	(The circuit operates as defined.)
Appearance	_____	(Work shows a neat and professional manner.)
Safety factor	_____	(Student used proper safety practices and operated tools and equipment safely.)

CAUTION

Be sure to de-energize the electrical circuit before disassembling the wiring booth.

---

6. What did you find was the most challenging part of this lab assignment?

_____

_____

_____

7. What is the most interesting or important thing that you learned during this lab assignment?

_____

_____

_____

National Electrical Code® Drill Work

Select the correct answer for the following questions, and indicate the article or section in the *NEC* in which the answer can be found.

1. What is the minimum number of 120-volt, 15-ampere general lighting circuits required of a two-story residence that has an outside dimension of 45 ft × 57 ft (13.72 m × 17.37 m)?

   A. 5 circuits

   B. 7 circuits   *NEC*: _____

   C. 9 circuits

   D. 11 circuits

2. What is the minimum rated service-disconnecting means allowed for a one-family dwelling?

   A.   100-ampere, 2-wire

   B.   100-ampere, 3-wire   *NEC:* _____

   C.   200-ampere, 2-wire

   D.   200-ampere, 3-wire

3. What is the maximum number of 10 AWG THHN copper conductors allowed in a 3¾ × 2 × 3½ masonry metal box?

   A.   5

   B.   7   *NEC:* _____

   C.   8

   D.   9

4. What AWG UF conductor is required for a 25-ampere load if the ambient temperature is 43°C and there are eight current-carrying conductors in the raceway?

   A.   10 AWG

   B.   8 AWG   *NEC:* _____

   C.   6 AWG

   D.   4 AWG

5. True or false? It is allowable to use solder-dependent connectors to connect grounding conductors to enclosures.

   A.   True

   B.   False   *NEC:* _____

6. Which of the following duty operations is defined as "a substantially constant load for an indefinitely long time"?

   A.   Continuous duty

   B.   Periodic duty   *NEC:* _____

   C.   Short-time duty

   D.   Varying duty

Two-Entrance Door Chime Wiring Lab

Name:	Date:	Grade:

Student Learning Outcomes

After completing this lab, students should be able to:

- Describe the theory of operation and required components of a single-entrance door chime system.

- Describe the theory of operation and required components of a two-entrance door chime system.

- Properly install the wiring and connect the required components associated with a single-entrance door chime system.

- Properly install the wiring and connect the required components associated with a two-entrance door chime system.

Equipment and Supplies

Two door chime push buttons

Class 2 chime transformer rated between 10 and 24 volts (5–30 VA)

(for bell transformers: rated between 6 and 10 volts [5–20 VA])

Residential 2-tone door chime

Octagonal metal junction box

Insulated 2-conductor wire (choose 18–22 AWG conductors)

Procedures

CAUTION
Because portions of this lab involve working with 120 vac, your complete attention is required, and your every action must be well thought out and safety conscious.

POWER TOOL CAUTION
During the course of this lab assignment, it may be necessary to use various power tools and stepladders. It is imperative that a thorough understanding of the operating procedures and safety concerns associated with these power tools and stepladders be achieved before using them.

1. In the space provided here, sketch the basic floor plan of your wiring booth, and draw the wiring diagram that is needed for a single-entrance door chime system. (Refer to Appendix I for an example floor plan.) The circuit should be laid out in the following manner:

    A. The door chime push button should be located at the front of the left-hand panel and 3 ft (900 mm) from the floor to the bottom of the push button.

    B. The residential door chime should be located at the rear of the left-hand panel and 5 ft (1.5 m) from the floor to the bottom of the chime and directly above the push button.

    C. The Class 2 transformer should be located overhead and in the center of the booth.

2. In the space provided here, and using the connection method described in step 1, draw the exact wiring that is involved in residential door chime system described in step 1. (Refer to Appendix II for an example wiring diagram.) Have the instructor verify that your diagram is drawn correctly and initial below.

CAUTION

It is important that the instructor check your wiring diagram before you begin to complete the electrical connections.

Instructor's initials: _____

3. Once the instructor has verified the validity of your wiring diagram, wire your assigned booth in accordance with the wiring diagram in step 2.

CAUTION

Do not work on circuits with power applied. Notify the instructor before energizing the circuit, and de-energize the power whenever possible.

---

4. Once you have completed the installation, have the instructor inspect and grade your wiring booth, using the following criteria:

Electrical integrity   _____   (Splices, terminations, and grounding were correctly done.)

NEC® compliance   _____   (Local codes should be included.)

Operational   _____   (The circuit operates as defined.)

Appearance   _____   (Work shows a neat and professional manner.)

Safety factor   _____   (Student used proper safety practices and operated tools and equipment safely.)

CAUTION

Be sure to de-energize the electrical circuit before modifying the wiring booth.

---

5. Modify the current door chime system to a two-entrance door chime system by adding a second push button. It should be located at the front of the right-hand panel, directly across from the first push button, and 3 ft (900 mm) from the floor to the bottom of the push button. In the space provided here, draw the exact wiring that is involved in this two-entrance door chime system. (Refer to Appendix II for an example wiring diagram.) Have the instructor verify that your diagram is drawn correctly and initial below.

CAUTION

It is important that the instructor check your wiring diagram before you begin to complete the electrical connections.

Instructor's initials: _____

6. Once the instructor has verified the validity of your wiring diagram, wire your assigned booth in accordance with the wiring diagram in step 5.

CAUTION

Do not work on circuits with power applied. Notify the instructor before energizing the circuit, and de-energize the power whenever possible.

---

7. Once you have completed the installation, have the instructor inspect and grade your wiring booth, using the following criteria:

Electrical integrity   _____   (Splices, terminations, and grounding were correctly done.)

NEC compliance   _____   (Local codes should be included.)

Operational   _____   (The circuit operates as defined.)

Appearance   _____   (Work shows a neat and professional manner.)

Safety factor   _____   (Student used proper safety practices and operated tools and equipment safely.)

CAUTION

Be sure to de-energize the electrical circuit before disassembling the wiring booth.

---

8. What did you find was the most challenging part of this lab assignment?

_____

_____

_____

9. What is the most interesting or important thing that you learned during this lab assignment?

_____

_____

_____

National Electrical Code® Drill Work

Select the correct answer for the questions listed below and indicate the article or section in the *NEC* in which the answer can be found.

1. A circuit breaker that is being used as a switch shall be installed so that the maximum height of center grip of the operating handle of the circuit breaker is not more than ___ above the working platform.
   A. 4.5 ft (1.37 m)
   B. 6 ft (1.8 m)   *NEC*: _____
   C. 6 ft 7 in. (2.01 m)
   D. 7 ft (2.1 m)

2. A 120/240-volt, 3-wire service-drop passing over a residential driveway shall have a minimum clearance from final grade of how many feet?
   A. 10 ft (3.0 m)
   B. 12 ft (3.5 m)   *NEC*: _____
   C. 15 ft (4.5 m)
   D. 18 ft (6 m)

3. What is the maximum ampacity of a 8 AWG THHN conductor when it is installed on the roof of a house that has an ambient temperature of 62°C?

   A. 26 amperes

   B. 32 amperes   *NEC*: _____

   C. 52 amperes

   D. 94 amperes

4. True or false? It is permitted to use a medium-drawn copper as the lead-in conductor as part of the receiving equipment of an antenna system if the span between the points of support is equal to 30 ft (10 m).

   A. True

   B. False   *NEC*: _____

5. All of the following statements relate to XHH conductor insulation, *except*:

   A. It is suitable for dry locations.

   B. It is moisture-resistant.   *NEC*: _____

   C. It is suitable for damp locations.

   D. It is flame-retardant.

6. What is the minimum headroom of working space above a 6.5 ft (2 m) panelboard?

   A. 5.5 ft (1.67 m)

   B. 6 ft (1.8 m)   *NEC*: _____

   C. 6.5 ft (2 m)

   D. 7 ft (2.1 m)

Home Security System Wiring Lab

Name:	Date:	Grade:

Student Learning Outcomes

After completing this lab, students should be able to:

- Describe the theory of operation and components of a home security system.

- Properly install the wiring and connect the components required for a home security system.

- Properly upgrade a home security system with additional input devices.

- Properly upgrade a home security system with additional output devices.

Equipment and Supplies

Two motion detectors with lights

Home security control panel

One normally open detection device

Three normally closed detection devices

Two normally open push buttons

Overhead lamp (output device)

Insulated 2-conductor wire (choose 18–22 AWG conductors)

Procedures

> **CAUTION**
> Because portions of this lab involve working with 120 vac, your complete attention is required, and your every action must be well thought out and safety conscious.

> **POWER TOOL CAUTION**
> During the course of this lab assignment, it may be necessary to use various power tools and stepladder. It is imperative that a thorough understanding of the operating procedures and safety concerns associated with these power tools and stepladders be achieved before using them.

1. In the space provided on the next page, sketch the basic floor plan of your wiring booth, and draw the wiring diagram associated with the home security system that is described here. (Refer to Appendix I for an example floor plan.) The circuit should be laid out in the following manner:

   A. Mount the home security control panel in the middle of the rear panel of the wiring booth so that the bottom of the control panel is 56 in. (1.42 m) above the finished floor.

   B. Mount a motion detector in the upper-front corner of the left-hand panel so that the motion detector is facing the rear panel of the wiring booth.

   C. Connect a light to the alarm output of the control panel, and mount it overhead and in the center of the wiring booth.

   D. Mount a normally open (N.O.) device at the front of the left-hand panel so that it is 4 ft (1.2 m) from the finished floor, and connect the N.O. device to the delayed response input of the control panel. (These types of contacts are used at the main entrances of a residence.)

   E. Mount a normally closed (N.C.) device at the front of the right-hand panel so that it is 4 ft (1.2 m) from the finished floor, and connect the N.C. device to the instant-response input of the control panel. (These types of contacts are used at the windows.)

   F. Mount a N.O. push button on the left side of the rear panel so that it is 4 ft (1.2 m) from the finished floor, and connect the N.O. push button to the panic input of the control panel. (These types of contacts are normally mounted in the bedroom.)

2. In the space provided here, and using the connection method described in step 1, draw the exact wiring that is involved in the basic home security system described in step 1. (Refer to Appendix II for an example wiring diagram.) Have the instructor verify that your diagram is drawn correctly and initial below.

> **CAUTION**
> It is important that the instructor check your wiring diagram before you begin to complete the electrical connections.

Instructor's initials: _____

3. Once the instructor has verified the validity of your wiring diagram, wire your assigned booth in accordance with the wiring diagram in step 2.

> **CAUTION**
> Do not work on circuits with power applied. Notify the instructor before energizing the circuit, and de-energize the power whenever possible.

4. Once you have completed the installation, have the instructor inspect and grade your wiring booth, using the following criteria:

Electrical integrity   _____   (Splices, terminations, and grounding were correctly done.)

NEC® compliance   _____   (Local codes should be included.)

Operational   _____   (The circuit operates as defined.)

Appearance   _____   (Work shows a neat and professional manner.)

Safety factor   _____   (Student used proper safety practices and operated tools and equipment safely.)

CAUTION

Be sure to de-energize the electrical circuit before modifying the wiring booth.

---

5. Modify the basic home security system in the following manner, and draw the exact wiring required for the specified modifications.

A. Add two more N.C. devices to the instant response input side of the control panel, and mount the N.C. devices at the middle and rear of the right-hand panel; both devices should be 4 ft (1.2 m) off the finished floor.

B. Add a second N.O. push button to the panic input of the control panel, and mount the N.O. push button on the front edge of the left-hand panel so that the push button is 4 ft (1.2 m) off the finished floor. (Refer to Appendix II for an example wiring diagram.) Have the instructor verify that your diagram is drawn correctly and initial below.

CAUTION

It is important that the instructor check your wiring diagram before you begin to complete the electrical connections.

Instructor's initials: _____

6. Once the instructor has verified the validity of your wiring diagram, wire your assigned booth in accordance with the wiring diagram in step 5.

CAUTION

Do not work on circuits with power applied. Notify the instructor before energizing the circuit, and de-energize the power whenever possible.

---

7. Once you have completed the installation, have the instructor inspect and grade your wiring booth, using the following criteria:

Electrical integrity   _____   (Splices, terminations, and grounding were correctly done.)

NEC compliance   _____   (Local codes should be included.)

Operational   _____   (The circuit operates as defined.)

Appearance   _____   (Work shows a neat and professional manner.)

Safety factor   _____   (Student used proper safety practices and operated tools and equipment safely.)

CAUTION

Be sure to de-energize the electrical circuit before disassembling the wiring booth.

---

8. What did you find was the most challenging part of this lab assignment?

_____

_____

_____

9. What is the most interesting or important thing that you learned during this lab assignment?

_____

_____

_____

National Electrical Code® Drill Work

Select the correct answer for the following questions, and indicate the article or section in the *NEC* in which the answer can be found.

1. Nine 12 AWG RHW copper conductors are to be enclosed on a conduit raceway. The conduit is in an area that has an ambient temperature of 93°F. What is the maximum ampacity of these conductors?

    A. 16.5 amperes

    B. 23.5 amperes   *NEC*: _____

    C. 26.6 amperes

    D. 38 amperes

2. Define the term *enclosed*.

_____

_____

_____

NEC: _____

3. When outdoor antennas are in the proximity of power service-entrance conductors of less than 250 volts, what is the minimum allowable clearance between the antenna and the service-entrance conductors?

   A.   1 ft (300 mm)

   B.   2 ft (600 mm)   *NEC:* _____

   C.   3 ft (900 mm)

   D.   4 ft (1.2 m)

4. What is the minimum size fixture wire allowed?

   A.   4 AWG

   B.   16 AWG   *NEC:* _____

   C.   18 AWG

   D.   20 AWG

5. Which of the optical fiber cables listed below is defined as a factory assembly of one or more optical fibers having an overall covering and containing no electrically conductive materials?

   A.   Nonconductive optical fiber cable

   B.   Abandoned optical fiber cable   *NEC:* _____

   C.   Conductive optical fiber cable

   D.   Composite optical fiber cable

6. What is the minimum size bonding jumper that shall be connected between the communications-grounding electrode and the power-grounding electrode system structure where separate grounding electrodes are used?

   A.   6

   B.   8   *NEC:* _____

   C.   10

   D.   12

Fire Alarm System Wiring Lab

Name:	Date:	Grade:

Student Learning Outcomes

After completing this lab, students should be able to:

- Describe the theory of operation and components of a fire alarm system.

- Properly install the wiring and connect the components required for a fire alarm system.

- Properly upgrade a fire alarm system with additional fire alarms.

Equipment and Supplies

Three residential smoke detectors

Three octagonal metal junction boxes

14/2 NMB cable

Cable staples

Procedures

> **CAUTION**
> Because portions of this lab involve working with 120 vac, your complete attention is required, and your every action must be well thought out and safety conscious.

> **POWER TOOL CAUTION**
> During the course of this lab assignment, it may be necessary to use various power tools and stepladders. It is imperative that a thorough understanding of the operating procedures and safety concerns associated with these power tools and stepladders be achieved before using them.

1. In the space provided on the top of the next page, sketch the basic floor plan of your wiring booth, and draw the wiring diagram that is needed for a residential fire alarm system. (Refer to Appendix I for an example floor plan.) The circuit should be laid out in the following manner:

   A. The first smoke detector should be located on the left-hand panel and 6 ft (1.8 m) from the floor to the bottom of the detector.

   B. The second smoke detector should be located on the right-hand panel and 6 ft (1.8 m) from the floor to the bottom of the detector.

2. In the space provided here, and using the connection method described in step 1, draw the exact wiring that will be involved in residential fire alarm system described in step 1. (Refer to Appendix II for an example wiring diagram.) Have the instructor verify that your diagram is drawn correctly and initial below.

CAUTION

It is important that the instructor check your wiring diagram before you begin to complete the electrical connections.

Instructor's initials: _____

3. Once the instructor has verified the validity of your wiring diagram, wire your assigned booth in accordance with the wiring diagram in step 2.

CAUTION

Do not work on circuits with power applied. Notify the instructor before energizing the circuit, and de-energize the power whenever possible.

---

4. Once you have completed the installation, have the instructor inspect and grade your wiring booth, using the following criteria:

Electrical integrity   _____   (Splices, terminations, and grounding were correctly done.)

NEC® compliance   _____   (Local codes should be included.)

Operational   _____   (The circuit operates as defined.)

Appearance   _____   (Work shows a neat and professional manner.)

Safety factor   _____   (Student used proper safety practices and operated tools and equipment safely.)

CAUTION

Be sure to de-energize the electrical circuit before modifying the wiring booth.

---

5. Modify the residential fire alarm system to a three-fire alarm by adding a third smoke detector, which should be located overhead and in the center of the booth. In the space provided here, draw the exact wiring that is involved in this modified residential fire alarm system. (Refer to Appendix II for an example wiring diagram.) Have the instructor verify that your diagram is drawn correctly and initial below.

CAUTION

It is important that the instructor check your wiring diagram before you begin to complete the electrical connections.

Instructor's initials: _____

6. Once the instructor has verified the validity of your wiring diagram, wire your assigned booth in accordance with the wiring diagram in step 5.

CAUTION

Do not work on circuits with power applied. Notify the instructor before energizing the circuit, and de-energize the power whenever possible.

---

7. Once you have completed the installation, have the instructor inspect and grade your wiring booth, using the following criteria:

Electrical integrity	_____	(Splices, terminations, and grounding were correctly done.)
NEC compliance	_____	(Local codes should be included.)
Operational	_____	(The circuit operates as defined.)
Appearance	_____	(Work shows a neat and professional manner.)
Safety factor	_____	(Student used proper safety practices and operated tools and equipment safely.)

CAUTION

Be sure to de-energize the electrical circuit before disassembling the wiring booth.

---

8. What did you find was the most challenging part of this lab assignment?

_____

_____

_____

9. What is the most interesting or important thing that you learned during this lab assignment?

_____

_____

_____

National Electrical Code® Drill Work

Select the correct answer for the questions listed below, and indicate the article or section in the *NEC* in which the answer can be found.

1. Excluding any exceptions, when power-limited fire alarm circuits are installed in hoistways the circuit must be installed in any of the following, *except*:

    A. Rigid metal conduit
    B. Intermediate metal conduit
    C. Electrical metallic tubing
    D. Rigid PVC, Schedule 40

    NEC: _____

2. Which of the following is defined as an intermittent operation in which the load conditions are regularly recurrent?

A. Continuous duty

B. Periodic duty   *NEC:* _____

C. Short-time duty

D. Varying duty

3. In a power-limited fire alarm circuit, the transformer has to be either a listed PLFA transformer or a ____ transformer.

A. Class 1

B. Class 2   *NEC:* _____

C. Class 3

D. None of the above

4. A non-power-limited fire alarm circuit made of 18 AWG conductor that requires an overcurrent protection shall not exceed how many amperes?

A. 7 amperes

B. 9 amperes   *NEC:* _____

C. 10 amperes

D. 11 amperes

5. For further information on the installation and monitoring for integrity requirements for fire alarm systems, the *NEC* recommends referring to which NFPA standard?

A. NFPA 36

B. NFPA 50   *NEC:* _____

C. NFPA 70

D. NFPA 72

6. All of the following metal boxes are capable of holding nine 14 AWG USE copper conductors, *except*:

A. 3 in. × 2 in. × 2½ in. metal device box

B. 3¾ in. × 2 in. × 3½ in. masonry metal box   *NEC:* _____

C. 4 in. × 1½ in. square metal box

D. 4 in. × 2⅛ in. octagonal metal box

Service Entrance Wiring Lab

Name:	Date:	Grade:

Student Learning Outcomes

After completing this lab, students should be able to:

- Identify the required components of residential service entrance.

- Explain the theory of operation of a residential service entrance.

- Properly mount and install the wiring and components associated with a residential service entrance.

Equipment and Supplies

Residential load center with a 100-ampere main breaker

Meter socket

Service head

20-ampere circuit breaker

15-ampere circuit breaker

Ground rod clamp

Three 1½ in. rigid conduit straps

Three aluminum split bolts for the 2 AWG aluminum conductor

6 ft of 6 AWG bare copper conductor

8 ft of 1½ in. rigid conduit

25 ft of 2 AWG Aluminum SE conductor

(Refer to local codes where appropriate.)

Procedures

> **CAUTION**
> Because portions of this lab involve working with high voltage, your complete attention is required, and your every action must be well thought out and safety conscious.

> **POWER TOOL CAUTION**
> During the course of this lab assignment, it may be necessary to use various power tools and stepladders. It is imperative that a thorough understanding of the operating procedures and safety concerns associated with these power tools and stepladders be achieved before using them.

1. In the space provided here, sketch the basic floor plan of your wiring booth. (Refer to Appendix I for an example floor plan.) Then draw the wiring diagram that is needed to install the service entrance to the booth in accordance with the following description:

   A. The meter socket should be placed on the front outside edge of the booth's left-hand panel and 5 ft (1.5 m) from the floor to the centerline of the meter socket.

   B. The residential load center, with a 100-ampere main breaker, should be placed at the front of the left-hand panel and 79 in. (2.01 m) from the floor to the top of the residential load center.

      The *NEC*® states that the disconnect handle in its highest position cannot be higher than 79 in. (2.01 m); therefore, you need to make sure that once your load center is mounted, this requirement is not being violated.

   C. The mast should be tall enough to raise the service head above the top of the booth by 20 in. (500 mm).

2. In the space provided here, and using the connection method described in step 1, draw the exact wiring that is associated with the residential service entrance for this booth. (Refer to Appendix II for an example wiring diagram.) Have the instructor verify that your diagram is drawn correctly and initial below.

CAUTION

It is important that the instructor check your wiring diagram before you begin to complete the electrical connections.

Instructor's initials: _____

3. Once the instructor has verified the validity of your wiring diagram, wire your assigned booth in accordance with the wiring diagram in step 4 (account for ½ in. [12.7 mm] thick gypsum board).

CAUTION

Do not work on circuits with power applied. Notify the instructor before energizing the circuit, and de-energize the power whenever possible.

---

4. Once you have completed the installation, have the instructor inspect and grade your wiring booth, using the following criteria:

Electrical integrity   _____   (Splices, terminations, and grounding were correctly done.)

NEC compliance   _____   (Local codes should be included.)

Operational   _____   (The circuit operates as defined.)

Appearance   _____   (Work shows a neat and professional manner.)

Safety factor   _____   (Student used proper safety practices and operated tools and equipment safely.)

CAUTION

Be sure to de-energize the electrical circuit before securing the wiring booth.

---

5. What did you find was the most challenging part of this lab assignment?

_____

_____

_____

6. What is the most interesting or important thing that you learned during this lab assignment?

_____

_____

_____

National Electrical Code® Drill Work

Select the correct answer for the following questions, and indicate the article or section in the *NEC* in which the answer can be found.

1. What is the minimum size copper-clad aluminum conductor that can be used for a service lateral *not* being used for supplying only limited loads of a single branch circuit?

   A.   6 AWG

   B.   8 AWG   *NEC*: _____

   C.   10 AWG

   D.   12 AWG

2. What is the minimum allowable size of a rigid metal conduit containing six 6 AWG THHW aluminum conductors?

   A. 1½ in.

   B. 2 in.   *NEC:* _____

   C. 2½ in.

   D. 3 in.

3. True or false? Service conductors supplying a building are allowed to pass through the interior of another building.

   A. True

   B. False   *NEC:* _____

4. All of the following articles have not been changed since the 2011 *NEC, except*:

   A. *300.7*

   B. *430.9*

   C. *356.20*

   D. *110.9*

5. Define the term *weatherproof.*

   _____

   _____

   _____

6. What is the minimum number of inches (millimeters) between the top surface of a finished grade and the top of a buried rigid metal conduit?

   A. 6 in. (150 mm)

   B. 12 in. (300 mm)   *NEC:* _____

   C. 18 in. (450 mm)

   D. 24 in. (600 mm)

Subpanel Wiring Lab

Name:	Date:	Grade:

Student Learning Outcomes

After completing this lab, students should be able to:

- Identify the required components of residential subpanel.

- Explain the theory of operation of a residential subpanel.

- Properly mount and install the wiring and components associated with a residential subpanel.

- Test subpanel assembly by connecting a simple single switched light circuit to the subpanel.

Equipment and Supplies

Residential subpanel

20-ampere circuit breaker

15-ampere circuit breaker

6 ft of 6 AWG bare copper conductor

25 ft of 2 AWG aluminum SE conductor

Single-pole switch, switch box

120-volt luminaire, lighting outlet

14/2 NMB cable

Cable staples

(Refer to local codes where appropriate.)

Procedures

CAUTION

Because portions of this lab involve working with high voltage, your complete attention is required, and your every action must be well thought out and safety conscious.

POWER TOOL CAUTION

During the course of this lab assignment, it may be necessary to use various power tools and stepladders. It is imperative that a thorough understanding of the operating procedures and safety concerns associated with these power tools and stepladders be achieved before using them.

1. In the space provided here, sketch the basic floor plan of your wiring booth, and draw the wiring diagram that is needed to install the subpanel to the booth in accordance with the following description: (Refer to Appendix I for an example floor plan.)

   A. The residential subpanel should be placed at the rear of the right-hand panel and 79 in. (2.01 m) from the floor to the top of the residential subpanel.

      The *NEC*® states that the disconnect handle in its highest position cannot be higher than 79 in. (2.01 m); therefore, you need to make sure that once your subpanel is mounted, this requirement is not being violated.

   B. The switch for the light should be placed at the front of the right-hand panel, and the 120-volt source of this light circuit should come from the subpanel.

   C. The luminaire should be located overhead and in the center of the booth.

2. In the space provided here, and using the connection method described in step 1, draw the exact wiring that is associated with the residential service entrance for this booth. (Refer to Appendix II for an example wiring diagram.) Have the instructor verify that your diagram is drawn correctly and initial below.

CAUTION

It is important that the instructor check your wiring diagram before you begin to complete the electrical connections.

Instructor's initials: _____

3. Once the instructor has verified the validity of your wiring diagram, wire your assigned booth in accordance with the wiring diagram in step 2 (account for ½ in. [12.7 mm] thick gypsum board).

> **CAUTION**
>
> Do not work on circuits with power applied. Notify the instructor before energizing the circuit, and de-energize the power whenever possible.

4. Once you have completed the installation, have the instructor inspect and grade your wiring booth, using the following criteria:

Electrical integrity	_____	(Splices, terminations, and grounding were correctly done.)
NEC compliance	_____	(Local codes should be included.)
Operational	_____	(The circuit operates as defined.)
Appearance	_____	(Work shows a neat and professional manner.)
Safety factor	_____	(Student used proper safety practices and operated tools and equipment safely.)

> **CAUTION**
>
> Be sure to de-energize the electrical circuit before securing the wiring booth.

5. What did you find was the most challenging part of this lab assignment?

_____

_____

_____

6. What is the most interesting or important thing that you learned during this lab assignment?

_____

_____

_____

National Electrical Code® Drill Work

Select the correct answer for the following questions, and indicate the article or section in the *NEC* in which the answer can be found.

1. What is the maximum number of sets of disconnects for each service grouped in any one location?
   A. 2
   B. 3
   C. 6
   D. 8

   NEC: _____

2. The following statements relate to THHN conductor insulation, *except*:
   A.   It is heat-resistant material.
   B.   It is suitable for damp locations.   *NEC:* _____
   C.   It is suitable for wet locations.
   D.   It is flame-retardant material.

3. If a raceway is longer than 24 in. (600 mm), what are the two factors that must be determined before derating the size of a conductor?
   A.   Dampness factor and ambient temperature
   B.   Ambient temperature and number of conductors   *NEC:* _____
   C.   Number of conductors and dampness factor

4. What is the minimum size electrical nonmetallic tubing?
   A.   ⅜ in. (9.53 mm)
   B.   ½ in. (12.7 mm)   *NEC:* _____
   C.   ¾ in. (19.05 mm)
   D.   1 in. (25 mm)

5. Service conductors installed as open conductors shall have a clearance of not less than _____ feet from windows that are designed to be opened.
   A.   2 ft (600 mm)
   B.   3 ft (900 mm)   *NEC:* _____
   C.   6 ft (1.8 m)
   D.   10 ft (3.0 m)

6. Excluding any exceptions, when 2 AWG copper service-entrance conductors are used, a copper grounding electrode cannot be less than:
   A.   2 AWG
   B.   4 AWG   *NEC:* _____
   C.   6 AWG
   D.   8 AWG

Receptacle Add-On Wiring Lab

Name:	Date:	Grade:

Student Learning Outcomes

After completing this lab, students should be able to:

- Perform the calculations involved in selecting the correctly sized outlet box for the assigned task.

- Properly mount outlets, install the wiring and connect the devices required by the following electrical circuits:

  - An overhead light circuit that is controlled at one location.
  - A duplex-receptacle circuit with two duplex receptacles.

  A 2-gang box is used to house the receptacle circuit switch and one of the switches of the light circuit.

- Demonstrate the skills needed to:

  - Convert the overhead light circuit so that it can be controlled at two locations.
  - Add an additional duplex receptacle to the previously installed receptacle circuit.

Equipment and Supplies

Single-pole switch, switch box

Two 3-way switches, switch boxes

Two duplex receptacles, receptacle outlets

120-volt luminaire, lighting outlet

14/2 NMB cable

14/3 NMB cable

Cable staples

Procedures

CAUTION
Because portions of this lab involve working with 120 vac, your complete attention is required, and your every action must be well thought out and safety conscious.

POWER TOOL CAUTION
During the course of this lab assignment, it may be necessary to use various power tools and stepladders. It is imperative that a thorough understanding of the operating procedures and safety concerns associated with these power tools and stepladders be achieved before using them.

117

1. In the space provided here, sketch the basic floor plan of your wiring booth, and draw the wiring diagram required to install the electrical circuits described here. (Refer to Appendix I for an example floor plan.)

A. An overhead light circuit that is controlled at one location and is to be laid out in the following manner:

1. The switch should be located at the front of the left-hand panel and 4 ft (1.2 m) from the floor to the bottom of the box, and the 120-volt source is to enter at this switch.

2. The luminaire should be located overhead and in the center of the booth.

B. A split duplex-receptacle circuit with two duplex receptacles that are controlled by a single switch.

1. The switch should be ganged with the light circuit switch and the 120-volt source should enter this circuit at this switch.

2. One duplex receptacle should be located directly below the light switch and 1 ft (300 mm) from the floor to the bottom of the box.

3. The other duplex receptacle should be located at the rear of the left-hand panel and 1 ft (300 mm) from the floor to the bottom of the box.

2. In the space provided here, and using the connection methods described in step 1, draw the exact wiring that is involved in the electrical circuits specified in step 1. (Refer to Appendix II for an example wiring diagram.) Have the instructor verify that your diagram is drawn correctly and initial below.

CAUTION
It is important that the instructor check your wiring diagram before you begin to complete the electrical connections.

Instructor's initials: _____

3. Once the instructor has verified the validity of your wiring diagram, wire your assigned booth in accordance with the wiring diagram in step 2.

CAUTION

Do not work on circuits with power applied. Notify the instructor before energizing the circuit, and de-energize the power whenever possible.

---

4. Once you have completed the installation, have the instructor inspect and grade your wiring booth, using the following criteria:

Electrical integrity   _____   (Splices, terminations, and grounding were correctly done.)

NEC® compliance   _____   (Local codes should be included.)

Operational   _____   (The circuit operates as defined.)

Appearance   _____   (Work shows a neat and professional manner.)

Safety factor   _____   (Student used proper safety practices and operated tools and equipment safely.)

CAUTION

Be sure to de-energize the electrical circuit before modifying the wiring.

---

5. Draw the electrical modifications required to convert the two previous circuits in the following manner:

A. Modify the overhead light circuit by converting it to a circuit that can be controlled at two locations. The additional switch should be located on the rear panel of the booth and 4 ft (1.2 m) from the floor to the bottom of the box.

B. An additional duplex receptacle should be added to the duplex-receptacle circuit, and the new outlet should be located on the rear panel of the booth and 1 ft (300 mm) from the floor to the bottom of the box.

   The alterations should be accomplished as if all of the panels of the booth were covered with gypsum board. (Refer to Appendix II for an example wiring diagram.) Have the instructor verify that your diagram is drawn correctly and initial below.

CAUTION

It is important that the instructor check your wiring diagram before you begin to complete the electrical connections.

Instructor's initials: _____

6. Once the instructor has verified the validity of your wiring diagram, wire your assigned booth in accordance with the wiring diagram in step 5.

CAUTION

Do not work on circuits with power applied. Notify the instructor before energizing the circuit, and de-energize the power whenever possible.

---

7. Once you have completed the installation, have the instructor inspect and grade your wiring booth, using the following criteria:

Electrical integrity   _____   (Splices, terminations, and grounding were correctly done.)

NEC compliance   _____   (Local codes should be included.)

Operational   _____   (The circuit operates as defined.)

Appearance   _____   (Work shows a neat and professional manner.)

Safety factor   _____   (Student used proper safety practices and operated tools and equipment safely.)

CAUTION

Be sure to de-energize the electrical circuit before disassembling the wiring booth.

---

8. What did you find was the most challenging part of this lab assignment?

_____

_____

_____

9. What is the most interesting or important thing that you learned during this lab assignment?

_____

_____

_____

National Electrical Code® Drill Work

Select the correct answer for the following questions, and indicate the article or section in the *NEC* in which the answer can be found.

1. What is the maximum ampacity of seven 6 AWG THW conductors installed in a 25 in. (635 mm) long raceway in an environment with an ambient temperature of 92°F?

   A.   43 amperes

   B.   61 amperes   *NEC*: _____

   C.   69 amperes

   D.   99 amperes

2. All of the following conductor insulation material is suitable for "Wet Locations," *except*:

   A.   MI
   B.   ZW   *NEC:* _____
   C.   THW
   D.   RHH

3. What is the minimum allowable size of an outdoor antenna conductor made of copper-clad steel that connects an outdoor antenna that is 50 ft (16 m) from the receiving station?

   A.   14 AWG
   B.   17 AWG   *NEC:* _____
   C.   19 AWG
   D.   20 AWG

4. The following articles have not been changed since the 2008 *NEC*, *except*:

   A.   *810.21*
   B.   *422.18*
   C.   *230.92*
   D.   *250.58*

5. What is the maximum interval between conductor supports when the concealed knob-and-tube wiring method is used?

   A.   1 ft (300 mm)
   B.   3 ft (900 mm)   *NEC:* _____
   C.   4.5 ft (1.35 m)
   D.   6 ft (1.8 m)

6. A 120/240-volt, 3-wire service-drop passing over a public street shall have a minimum clearance from the final grade of how many feet?

   A.   10 ft (3.0 m)
   B.   12 ft (3.6 m)   *NEC:* _____
   C.   15 ft (4.5 m)
   D.   18 ft (5.5 m)

Post-Course Knowledge Assessment

Name:	Date:	Grade:

Student Learning Outcomes

After completing this lab, students should be able to:

- State which components and hardware of residential electricity they currently understand.

- Recognize which components and hardware of residential electricity they need to master before the conclusion of this course.

- State which residential electricity tools and equipment they currently understand.

- Recognize which residential electricity tools and equipment they need to master before the conclusion of this course.

Equipment and Supplies

The instructor secures 25–30 commonly used residential wiring parts and components and 10–15 commonly used hand tools and power tools that a practicing electrician needs to know how to use.

Procedures

Assessment of Prior Knowledge of Residential Wiring Parts and Components

1. The instructor has laid out 25–30 commonly used residential parts and components, which you will try to identify.

2. Go around to each part and, to the best of your ability, identify the name of the electrical component, and write a brief description of what or how the component is used in residential wiring.

Part 1: _____ : _____

_____

Part 2: _____ : _____

_____

Part 3: _____ : _____

_____

Part 4: _____ : _____

_____

Part 5: _____ : _____

_____

Part 6: _____ : _____

_____

Part 7: _____ : _____

_____

Part 8: _____ : _____

_____

Part 9: _____ : _____

_____

Part 10: _____ : _____

_____

Part 11: _____ : _____

_____

Part 12: _____ : _____

_____

Part 13: _____ : _____

_____

Part 14: _____ : _____

_____

Part 15: _____ : _____

_____

Part 16: _____ : _____

_____

Part 17: _____ : _____

_____

Part 18: _____ : _____

_____

Part 19: _____ : _____

_____

Part 20: _____ : _____

_____

Part 21: _____ : _____

_____

Part 22: _____ : _____

_____

Part 23: _____ : _____

_____

Part 24: _____ : _____

_____

Part 25: _____ : _____

_____

Part 26: _____ : _____

_____

Part 27: _____ : _____

_____

Part 28: _____ : _____

_____

Part 29: _____ : _____

_____

Part 30: _____ : _____

_____

Assessment of Prior Knowledge of Residential Wiring Hand and Power Tools

   3. The instructor has laid out 10–15 commonly used residential wiring hand and power tools, which you will try to identify.

   4. Identify the name of each tool, and write a brief description of how the tool is used in residential wiring.

Tool 1: _____ : _____

_____

Tool 2: _____ : _____

_____

Tool 3: _____ : _____

_____

Tool 4: _____ : _____

_____

Tool 5: _____ : _____

_____

Tool 6: _____ : _____

_____

Tool 7: _____ : _____

_____

Tool 8: _____ : _____

_____

Tool 9: _____ : _____

_____

Tool 10: _____ : _____

_____

Tool 11: _____ : _____

_____

Tool 12: _____ : _____

_____

Tool 13: _____ : _____

_____

Tool 14: _____ : _____

_____

Tool 15: _____ : _____

_____

5. How would you personally rate your progress in this course?

_____

_____

6. What was the most interesting or important thing that you learned during this lab assignment?

_____

_____

National Electrical Code® Drill Work

Select the correct answer for the questions listed below, and indicate the article or section in the *NEC*® in which the answer can be found.

1. What is the minimum distance an open individual (aerial) overhead conductor can be to any building or structure, other than supporting poles or towers, before it is required to be insulated or covered?
   A. 2 ft (600 mm)
   B. 3 ft (900 mm)   *NEC*: _____
   C. 6 ft (1.8 m)
   D. 10 ft (3.0 m)

2. Provide the *NEC*'s definition of a *nonmetallic-sheathed cable*.

   _____

   _____

                                   NEC: _____

3. True or false? It is permissible to use the junction of tree branches as support of overhead service conductors as long as the conductors meet the height regulations in terms of walkways and driveways.
   A. True
   B. False   *NEC*: _____

4. Receptacles and cord connectors shall be of a type that is unsuited for use as lampholders and shall be rated not less than:

    A.   15 amperes, 125 volts or 15 amperes, 250 volts

    B.   15 amperes, 125 volts or 20 amperes, 250 volts   *NEC:* _____

    C.   20 amperes, 125 volts or 15 amperes, 250 volts

    D.   20 amperes, 125 volts or 20 amperes, 250 volts

5. Nonmetallic underground conduit with conductors is permitted to be used in the following, *except*:

    A.   Encased or embedded in concrete

    B.   In exposed locations   *NEC:* _____

    C.   In cinder fill

    D.   For direct burial underground installation

6. True or false? It is permissible to attach a 9 lb. (4.5 kg) luminaire to a box that is secured to another box, as long as two No. 6 screws are used to secure the luminaire or its yoke to the box.

    A.   True

    B.   False   *NEC:* _____

Example Floor Plan Diagram

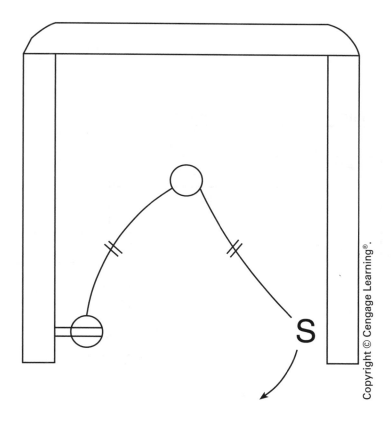

Example Wiring Diagram

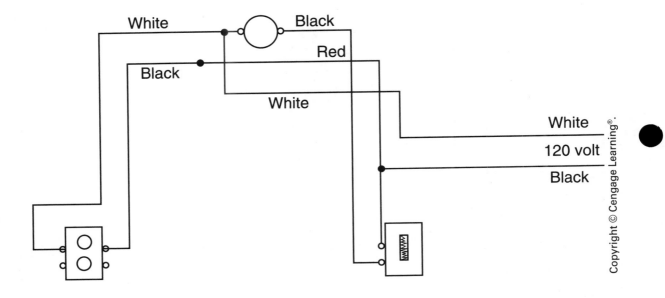

White

Black

Red

Black

White

White

Black

120 volt

APPENDIX III

Recommended Tool List

Tool pouch

Tool pouch belt

Hammer, 12 in. (300 mm) handle

Lineman pliers, 9 in. (229 mm)

Needlenose pliers, 8 in. (200 mm) or larger

Stakon pliers

Straight-slot screwdriver, ¼ in. × 6 in. (6.355 mm × 153 mm) shank

Straight-slot screwdriver, 3/16 in. × 4 in. (4.767 mm × 102 mm) shank

Straight-slot screwdriver, 5/16 in. × 5 in. (7.94 mm × 127 mm) shank

Phillips screwdriver, no. 2

Wire stripper/cutters

Automatic wire strippers

Electrician's knife

Electrical tape, black

Electrical tape, red

Measuring tape, 25 ft (7.5 m)

Safety glasses/goggles

Multimeter

The instructor is responsible for determining what part of this tool list will be provided by the training institute.

Materials List

Resistors:	2.2 kΩ, 4.7 kΩ, 10 kΩ, 15 kΩ (all 1 W resistors)

Conductors:	10 AWG insulated solid
	12 AWG insulated solid
	14 AWG insulated solid
	14/2 NMB cable

10 AWG insulated stranded

12 AWG insulated stranded

16 AWG insulated solid

14/3 NMB cable

Insulated 2-conductor wire (between 18 and 22 AWG)

6 AWG bare copper

2 AWG aluminum SE

Nonmetallic boxes:

3 in. × 2 in. × 2 in. device (2)   3 in. × 2 in. × 2¼ in. device

3 in. × 2 in. × 2½ in. device (3)   3 in. × 2 in. × 2¾ in. device (3)

3 in. × 2 in. × 3½ in. device (4)

4 in. × 1¼ in. round (2)   4 in. × 1½ in. round

4 in. × 2⅛ in. round   22.5-in.^{3} capacity round

Two-ganged   Three-ganged

Octagonal metal junction box (3)

Nonmetallic box extension, round

Switches:	Single-pole (4)	Three-way (2)
	Four-way (2)	

Receptacles:	Duplex receptacle (5)	GFCI duplex receptacle

120-volt luminaire (2)

Twist-on wire connectors (red, yellow, and green, grounding with pigtail)

Solderless terminal connectors (10 AWG, 12 AWG, and 14 AWG)

Crimp sleeve connectors

Residential 2-tone door chime

Door chime push buttons (2)

Class 2 chime transformer (rated 10–24 volts, 5–30 VA)

Motion detectors with lights (2)

Home security control panel

Normally open detection devices

Normally closed detection devices (3)

Normally open push buttons (2)

Residential smoke detectors (3)

Residential load center with a 100-ampere main breaker

Residential subpanel

Meter socket

Service head

15-ampere circuit breakers (4)

20-ampere circuit breakers (3)

Ground rod clamp

1½ in. rigid conduit straps (3)

Aluminum split bolt, for the 2 AWG aluminum conductor (3)

1½ in. rigid conduit

Electrical Symbols

Symbol	Description	Symbol	Description
S	Alarm—smoke	J	Junction box—ceiling
H	Alarm—heat	J	Junction box—wall
⊣I∣I∣⊢	Battery	▭ or ▭	Lighting or power panel, recessed
(Buzzer symbol)	Buzzer	▭ or ▭	Lighting or power panel, surface
(Circuit breaker symbol)	Circuit breaker	MD	Motion detector
▽	Data outlet	M	Motor
xxAF / yyAT	Disconnect switch, fused; size as indicated on drawings. "xxAF" indicates fuse ampere rating. "yyAT" indicates switch ampere rating.	2	Motor: "2" indicates horsepower
		(Overload relay symbol)	Overload relay
xxA	Disconnect switch, unfused; size as indicated on drawings. "xxA" indicates switch ampere rating.	⊡	Push button
(Doorbell symbol)	Doorbell	(Switch and fuse symbol)	Switch and fuse
CH	Door chime	▼ or ▼	Telephone outlet
D	Door opener (electric)	▼W or ▼W	Telephone outlet—wall mounted
(Fan paddle symbol)	Fan: ceiling-suspended (paddle)	▽	Telephone/data outlet
(Fan paddle with light symbol)	Fan: ceiling-suspended (paddle) fan with light	T L	Thermostat—line voltage
		T LV	Thermostat—low voltage
(Fan wall symbol)	Fan: wall	TS	Time switch
(Ground symbol)	Ground	T	Transformer

OUTLETS	CEILING			WALL		
Surface-mounted incandescent	○	⊕	⊗	⊢○	⊢⊕	⊢⊗
Lampholder with pull switch	○ PS	Ⓢ		⊢○ PS	⊢Ⓢ	
Recessed incandescent	⊡	Ⓡ	⊘	⊢⊡	⊢Ⓡ	⊢⊘
Surface-mounted fluorescent	▭	▭○▭		⊓	⊓○⊓	
Recessed fluorescent	▱	▭○R▭		▱	▭○R▭	
Surface or pendant continuous row fluorescent	▭▭▭ ○▭▭▭					
Recessed continuous row fluorescent	▱▱▱ ○R▱▱▱					
Bare lamp fluorescent strip	⊢—╂—╂—⊣					
Surface or pendant exit	Ⓧ			⊸Ⓧ		
Recessed ceiling exit	Ⓡ Ⓧ			⊸Ⓡ Ⓧ		
Blanked outlet	Ⓑ			⊸Ⓑ		
Outlet controlled by low-voltage switching when relay is installed in outlet box	Ⓛ			⊸Ⓛ		
Junction box	Ⓙ			⊸Ⓙ		

RECEPTACLE OUTLETS

⊖	Single receptacle outlet	⊖ D	Clothes dryer outlet
⊖	Duplex receptacle outlet	◐	Exhaust fan outlet
⊕	Triplex receptacle outlet	ⓒ	Clock outlet
⊖	Duplex receptacle outlet, split wired	⊙	Floor outlet
⊕	Double duplex receptacle (quadplex)	⊖ X"	Multioutlet assembly; arrow shows limit of installation. Appropriate symbol indicates type of outlet. Spacing of outlets indicated by "X" inches.
⊖ WP	Weatherproof receptacle outlet	⊡ S	Floor single receptacle outlet F = flush mtd, S = surface mtd
⊖ GFCI	Ground-fault circuit interrupter receptacle outlet	⊡ F	Floor duplex receptacle outlet F = flush mtd, S = surface mtd
⊖ R	Range outlet	◣ S	Floor special-purpose outlet F = flush mtd, S = surface mtd

▲ DW — Special-purpose outlet (subscript letters indicate special variations: DW = dishwasher; also a, b, c, d, etc., are letters keyed to explanation on drawings or in specifications).

Courtesy of National Electrical Contractors Association.

SWITCH SYMBOLS

S	Single-pole switch
S_2	Double-pole switch
S_3	Three-way switch
S_4	Four-way switch
S_D	Door switch
S_{DS}	Dimmer switch
S_G	Glow switch toggle— glows in off position
S_K	Key-operated switch
S_{KP}	Key switch with pilot light
S_{LV}	Low-voltage switch
S_{LM}	Low-voltage master switch
S_{MC}	Momentary-contact switch
⟨M⟩	Occupancy sensor—wall mounted with "Off-auto" override switch
⟨M⟩P	Occupancy sensor—ceiling mounted; "P" indicates multiple switches wire-in parallel
S_P	Switch with pilot light on when switch is on
S_T	Timer switch
S_R	Variable-speed switch
S_{WP}	Weatherproof switch

CIRCUITING

A 1-2 ///→ Home run to panel. Number of arrows indicates number of circuits. Letters and numbers indicate circuit(s). Full slashes indicate ungrounded "hot" (or switch leg) circuit conductors. Half slashes indicate grounded neutral circuit conductor(s). No slashes indicate one "hot" and one neutral conductor.

———///●● Branch circuit. Full slashes indicate ungrounded "hot" (or switch leg) circuit conductors. Half slashes indicate grounded neutral circuit conductor(s). No slashes indicate one "hot" and one neutral conductor. Dots indicate equipment grounding conductors.

———————— Wiring concealed in construction in finished areas, exposed in unfinished areas

– – – – – – – Conduit concealed in or under floor slab

————————○ Conduit turning up

————————● Conduit turning down

————————
CO Conduit only (empty)

⌒ Switch leg; connects switched outlets with control points

Courtesy of National Electrical Contractors Association.